TRANSCENDING
THE SPEED OF
LIGHT

"How is it possible for consciousness to exist in the physical universe? This is the classic mind-body problem that has eluded philosophers for many generations. Now it appears that answers are within reach. The depth of Marc Seifer's scholarship and the clarity of his thinking make this book a worthwhile read for anyone interested in the frontiers of consciousness research."

JEFFREY MISHLOVE, PH.D.,
DEAN OF CONSCIOUSNESS STUDIES RESEARCH,
UNIVERSITY OF PHILOSOPHICAL RESEARCH

"This book by Marc Seifer is truly a tremendous work! It represents a remarkable accomplishment of gathering an enormous amount of relevant material and taking the reader through a lifetime of meticulous research! I know of no other book that does all of that so thoroughly, and I highly recommend this book to any and all readers who are seriously interested in the puzzling problem of the nature of mind and consciousness. Marc's work is an epochal achievement that will offer new thoughts to the reader for many decades to come."

COL. TOM BEARDEN (RETIRED),
AUTHOR OF EXCALIBUR BRIEFING

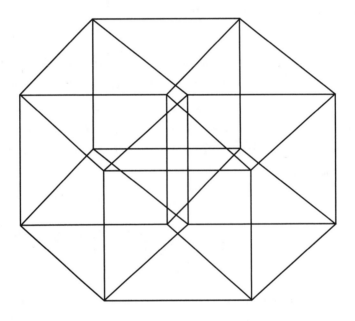

TRANSCENDING
THE SPEED OF
LIGHT

*CONSCIOUSNESS, QUANTUM PHYSICS,
AND THE FIFTH DIMENSION*

Marc Seifer, Ph.D.

Inner Traditions
Rochester, Vermont

Inner Traditions
One Park Street
Rochester, Vermont 05767
www.InnerTraditions.com

Library of Congress Cataloging-in-Publication Data
Seifer, Marc J.
  Transcending the speed of light : consciousness, quantum physics, and the fifth dimension / Marc Seifer.
    p. cm.
  Includes bibliographical references and index.
  ISBN 978-1-59477-229-0 (pbk.)
  1. Parapsychology. 2. Consciousness. 3. Physics—Miscellanea. I. Title.
  BF1031.S43 2008
  130—dc22

                                                                    2008010108

Printed and bound in the United States by Lake Book Manufacturing

10 9 8 7 6 5 4 3 2 1

Text design and layout by Jon Desautels
This book was typeset in Garamond Premier Pro with ITC Galliard as the display typeface

To send correspondence to the author of this book, mail a first-class letter to the author c/o Inner Traditions • Bear & Company, One Park Street, Rochester, VT 05767, and we will forward the communication.

For Lois, the light of my life

CONTENTS

FOREWORD

I first met Marc Seifer in the 1970s at a parapsychology conference in Washington, D.C. At that time, I had just finished a decade of conducting experiments with anomalous dreams at Brooklyn's Maimonides Medical Center Dream Laboratory. In fact, Marc told me that after he received his master's degree, he had visited Maimonides with the hope of apprenticing with me, but he found that I had moved to California, where I had begun teaching at Saybrook Graduate School.

A few years later, Marc enrolled at Saybrook, and I became his doctoral mentor. At that time, in the early 1980s, Marc conducted several independent studies on such topics as synchronicity and precognition. Even though neither of us knew it at the time, that work planted the seeds for this book.

As I have suggested in my autobiography, *Song of the Siren,* investigating this field is a tricky and hazardous matter because there are many blind alleys, several unknown variables, and no hard-and-fast rules as to what will make an ultimate contribution to human knowledge.

In *Dream Telepathy,* Montague Ullman, Alan Vaughan, and I presented the results of the controlled laboratory experiments we had conducted, which presented compelling evidence suggesting that some type of thought transference can occur while people are dreaming. Although other researchers amassed additional data, the serious study of such topics as telepathy in academic and medical environments remains virtually nonexistent, over three decades later. It is against this backdrop that Marc has continued his valiant work, performing his own studies in seeking to understand humanity's unknown capacities, and, beyond that, attempting to develop an overarching paradigm to explain these anomalous, puzzling phenomena.

My colleagues in parapsychological circles and I are endeavoring to further the field of research into human consciousness and to bring the results of our quest to the attention of the mainstream academic and medical communities. For the most part, these communities have yet to take such topics as telepathy seriously. One common question is, "But how do you explain your results? What are the mechanisms for telepathy and the other phenomena you are studying?" Marc has attempted to provide some answers to such questions. He is not trying to demonstrate that anomalous phenomena exist. A plethora of books and journal articles make this case. Instead, he cites anecdotes from his own life that illustrate putative synchronicity, telepathy, and precognition. These experiences will help many readers pay closer attention to exploring their own life experiences for possible instances of baffling phenomena that are difficult to explain in conventional terms.

Taking his cues from René Descartes, who began his speculations with a premise of doubting everything, and Thomas Kuhn, who suggested that scientific progress rests on the explanation of anomalies, Marc reexamines Descartes' mind/body dualistic paradigm and Einstein's supposition that nothing can travel faster than the speed of light. The mind for Marc is part of this physical world, a world that transcends the dualistic paradigm. Yet, paradoxically, it also has a transcendent function, and thus may inhabit a realm in which the speed of light has little relevance. Like many other writers, Marc calls this realm "inner space" or "hyperspace," although he gives the term his own unique spin.

To understand this realm, Marc calls for a program that will combine aspects of physics and psychology. His goal appears to be multifaceted. Not only is he attempting to provide his own description of the term "consciousness" in a way that portrays aspects of mind embedded in the structure of matter, but he also seeks to question some of the basic tenets of quantum physics. For example, he wants to reintroduce the long-discarded concept of "ether" to provide answers to some unexplained aspects of gravity, the spin of elementary particles, and, needless to say, psychic phenomena.

Basing his case on a wide variety of sources, Marc suggests that for physicists to produce the long-sought "grand unification theory" they must, by necessity, include the mind of the observer and the very process of consciousness itself. In Marc's paradigm, consciousness is a force comparable to gravity, electromagnetism, and the strong and weak nuclear forces.

I have my reservations about some aspects of this treatise, but not with its goal, namely to set the stage for establishing a paradigm for integrating consciousness into the structure of the spacetime continuum. Volition, intention, and expectancy play a role in experiments conducted by both psychologists and physicists; perhaps they play a role in weaving the fabric of reality itself.

STANLEY KRIPPNER, PH.D.

Stanley Krippner is past president of the Association for Humanistic Psychology and coauthor of the watershed book *Dream Telepathy: Experiments in Nocturnal Extrasensory Perception.* A professor of psychology at Saybrook Graduate School in San Francisco, Krippner is internationally known for his pioneering work in the investigation of human consciousness, parapsychological phenomena, and altered states of consciousness. He has written hundreds of articles and numerous books including *The Mythic Path, Becoming Psychic,* and *Healing States: A Journey into the World of Spiritual Healing and Shamanism.* Krippner has conducted workshops and seminars worldwide on dreams, hypnosis, and personal mythology.

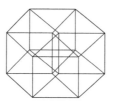

Preface

While working on the galleys of *Transcending the Speed of Light* and trying to locate a few missing references, I went into my archives to dig out my bulging file from my colleague Edwin Gora, professor emeritus at Providence College in Rhode Island, for it was Professor Gora who gave me so many articles that helped shape the nature and texture of this work. I first met Professor Gora in 1977, while I was teaching a series of courses on consciousness research at Providence College's night school. I was still in my twenties and he was, at least in my perspective, essentially an old man in his seventies, walking to the class on crutches. He had just had two hip replacements.

I was lecturing about Jung's theory of synchronicity, or meaningful coincidence, when Gora, sitting in the front row, raised his hand. "May I come up and say something?" He spoke in a distinct German accent. "Sure," I said. And Edwin moved to the blackboard and began to discuss cosmological coincidences (which appear in the first chapter), along with the theories of Carl Jung and his physics friend Wolfgang Pauli. Edwin became so engaged in what he was writing and saying, and frustrated from his encumbrances, that he threw his crutches to the ground and continued lecturing as he scratched out a half-dozen equations and then explained them to the class. We began our friendship that night.

I came to find that Professor Gora was half Polish and half German and that he had obtained his doctorate in physics at the height of World War II in Germany. His doctoral mentor was none other than Werner Heisenberg (from whom he had several handwritten letters), and he also knew Arnold Sommerfeld. As a graduate student in theoretical physics at the University of Munich, he knew nothing about Heisenberg's secret work on atomic weapons, and in fact Gora himself was picked up by the Gestapo because he was half Polish, and also because he had traveled to India and actually had a meeting with Mahatma Gandhi. Heisenberg helped save him, and both of them would play the game of espousing some of the Nazi rhetoric, even signing letters to each other with the closing *Heil Hitler!* Earlier, before the war, as Gora told me, the Nazi papers began to condemn Heisenberg because he taught Jewish physics. This led to Heisenberg himself being picked up and questioned. As luck would have it, Heisenberg's mother was friendly with Himmler's mother, and thus he was protected and thereby also allowed to teach the theory of relativity at the university. Gora became his last graduate student; Edward Teller (the father of the American A-bomb) had been his first.

Nearly thirty years later, Gora began an article for *MetaScience Quarterly,* a journal I was producing, with the title "Pythagorean Trends in Modern Physics." It took him about five years to complete the piece, with me editing every revision, prodding him on, with my mind spinning from our numerous meetings. I was not a physicist, so one of the goals was to write it so that I could understand it. If we could succeed there, then, it was hoped, our readership would understand it as well.

After we published the article; he began Part II. This was a project that lasted well over a decade. I kept prompting Edwin to complete it, but he could not because his actual goal was to solve the great cosmological question: How was the universe born and constructed, and where does consciousness fit in? I would meet him at his office, at Providence College, at College Hill Bookstore, where we would often run into each other, and during the summers, at the beach, because we both belonged to the same beach club; but he kept never finishing.

Along the way, however, Edwin (much like my mother) would send me one article after another on the great quest. Thumbing through them right now, I look at such titles as "When the Quarks Come Marching Home, Again," "Beyond Einstein: The Cosmic Quest for the Theory of the Universe," "Three Scientists and Their Gods," "Wormholes Might Open a Door to Other Dimensions [including] Time Travel," "From Chaos to Consciousness," and "The Unfinished Universe." The last one was a series of culled chapters or sections of chapters from Louise B. Young's book of the same title.

As with most of these abstracts, the Louise B. Young segment is filled with profound insights. "Although mankind appears to be just a minute local phenomenon in a cosmos so vast that its size humbles the imagination, size alone is not a measure of importance. We have seen that the transformation process takes place by building from tiny individual centers. The whole is immanent in all the parts, no matter how small."[1]

Professor Gora also included a cartoon of a man floating in a courtroom facing a judge who was admonishing him, saying, "You have broken the law of gravity. How do you plead?"

In 1993, Edwin wrote me shortly after his return from Munich, where he was teaching a course in astronomy. "I started writing something on 'God, Platonism, and the Universe,'" he wrote, "trying to bridge the views of Heisenberg v. Weizsacker, et al., with those of the current British 'Platonists' Barrow, Davies, and Penrose (see enclosure)." He died the following year shortly after translating, directly from the Greek, an ancient passage related to these concepts. He was eighty-two. Edwin, of course, had set himself up. Trying to solve the ultimate secret of the universe, for a cosmologist, was, to use one of his favorite words, "obviously" an impossible task. Thus he could never complete his treatise. My goals are not as lofty.

This book is the record of a quest: both a personal quest that I began in earnest in the early 1970s and the human quest to more fully grasp the nature of reality and the participation of human consciousness in it. The quest is not over and this book does not claim to be a final answer. What it does represent is some significant steps along the way and some pointers toward the directions that need to be explored more fully, bringing together insights from the realms of modern physics with those gleaned from the fields of psychology and consciousness research.

Transcending the Speed of Light, which originally had the subtitle *From Einstein to Ouspensky,* is really the second part of a two-part treatise, the first being *Inward Journey: From Freud to Gurdjieff* (2003). Although this text stands on its own, it is assumed that the reader has a working knowledge of the theories of Freud and Jung on the structure of the unconscious, works that are too often overlooked by traditional physicists in their attempts to create a model of the universe that takes cerebral processes into account.

To some extent, the book is an extension of my master's thesis, *Levels of Mind* (1974), which went beyond Freud and Jung into the realm of consciousness research, a topic I taught for fifteen years at Providence College's night school. A number of chapters were originally articles written for either *MetaScience Quarterly* or *Parapsychology Review,* including a review of salient and currently rare parapsychology texts written between 1873 and 1925 and reports on two significant symposiums of the late 1970s: one on the "Physics of Consciousness," which took place at the Harvard Science Center in 1977, and the other on the "Coevolution of Science and Spirit" in New York City in 1979. Although penned over a quarter century ago, the people and topics covered are still current because they deal with underlying concepts that must be addressed if we are to truly come to a comprehensive model that takes into account mind, time, and the fundamental structure of space.

Key ideas include the necessity for adding the dimension of inner or hyperspace to the structure of the universe to accommodate the mind; the reintroduction of serious consideration of the ether, which is at least a monadic sea of photons that incorporates the All; the idea that the mind already operates in tachyonic (that is, exceeding the speed of light) dimensions; and philosopher/mathematician P. D. Ouspensky's ideas on multidimensional time.

The sections on synchronicity and precognition were first penned as part of a special study I did for my doctoral mentor, Stanley Krippner, past president of the Association for Humanistic Psychology and coauthor of *Dream Telepathy,* a book that established conclusively that the REM cycle can be a channel for telepathic communication. Also included is a short lesson on astrology, which contains a specific method for predicting the future that is intimately linked to the structure of time. This section is based on courses I took with Zoltan Mason, an amazing astrologer who taught classes in New York City for many years. The book concludes with a fresh look at $E = mc^2$.

This book is the first edition of a controversial work. It tackles hard questions and paradoxical issues, giving consideration, for example, to both the traditional view of gravity and the etheric view; it sometimes regards the speed of light squared as a mathematical conversion factor and at other times sees it as operating in a new dimension. Criticisms and counterhypotheses are welcome, and the possibility exists that such comments may be integrated into future editions.

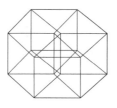

ACKNOWLEDGMENTS

This book would not have been published without the support of Inner Traditions. I would like to thank Ehud Sperling, Jeanie Levitan, Mindy Branstetter, Nancy Yeilding, and the rest of the ITI staff. I would also like to thank: Daniel G. Freedman, my master's thesis mentor at the University of Chicago; Roger Pearson, the dean at Providence College, for encouraging me; Howard Smukler, the first director of the MetaScience Foundation; Professor Stanley Krippner, who oversaw some of my graduate studies on such topics as synchronicity and precognition; Elliott Shriftman, Sandy Neuschatz, Godfrey Jordan, Lynn Sevigny, Tom Bearden, Ron Hatch, John White, Uri Geller, my brother and sister Bruce Seifer and Meri Shardin, and Edwin Gora, professor emeritus, for giving me numerous articles on the quantum physics of consciousness; and my father, Stanley Seifer, and my mother, Thelma Imber Seifer, for sending me numerous cogent articles related to this topic.

The book has been heavily influenced by such writers and thinkers as G. W. Leibniz, Albert Einstein, Hermann Minkowski, Arnold Sommerfeld, Charles Musès, Nikola Tesla, Arthur Koestler, David Bohm, Rudolf Steiner, Lobsang Rampa, Andrija Puharich, Uri Geller, Henri Bergson, John Ott, Royal Rife, Sigmund Freud, Alfred Landis, Paul Kammerer, and Carl Jung. Some important inspirations stem from the work of Dennis Gabor, the discoverer and developer of holography/3-D photography, and such books as George Gamow's *Thirty Years That Shook Physics,* Fritjof Capra's *The Tao of Physics,* books by Lincoln Barnett and James Coleman that explain Einstein's theories, and the amazing P. D. Ouspensky and his masterwork, *New Model of the Universe.*

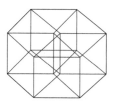

1

CONSCIOUSNESS AND THE ANTHROPIC PRINCIPLE

If the Universe is a product of mind . . . then it will ultimately illustrate mind's axiom.

J. W. DUNNE, 1934

The word *anthropic* refers to human beings. The term *anthropic principle* was introduced in 1973 by Brandon Carter, an astrophysicist from Cambridge, at a conference in Poland commemorating the five hundredth birthday of Copernicus, where Carter delivered a paper entitled "Large Number Coincidences and the Anthropic Principle in Cosmology." Carter suggested that highly specific details of the construction of the universe were necessary to allow "the emergence of observers at some stage." He noted, for example, that had the strong nuclear force—the force that holds the nucleus of an atom together—been just a "little stronger, protons would fail to form—a little weaker and the formation of stars would be impossible."[1] In other words, it had to be exactly as it was or life could not have evolved.

This correspondence between the precise structure of the universe and the emergence of carbon-based life systems that, at the top of the chain, resulted in entities that could think was seen by atheistic scientists as a coincidence, by Heinz Pagels as a form of "cosmic narcissism,"[2] and by scientists with a theological bent as evidence of a design-maker. Early theoreticians Johannes Kepler (1571–1630), Robert Boyle (1627–1691), and Richard Bentley (1662–1742) argued that the elegant mathematical basis of the movement of the planets was proof in and of itself of a design-maker. Isaac Newton (1642–1727), in discussing his discovery of the gravitational constant (which derived from Kepler's law of planetary motion), wrote, "Whence arises this uniformity in all their outward shapes, but from the counsel and contrivance of an Author."

Where Kepler "alluded to the Stoic idea that the universe was a living, rational, evolving being," René Descartes (1596–1650) conceived of this great design as something separate from the great design-maker. God's properties of purpose and thought reemerged only in the human soul, and not in other animals or anywhere else, according to Descartes' view.[3] Thus, he generated what became the dominating scientific paradigm of a world of consciousness split off from the physical mechanical-like universe. This model, essentially embraced by most modern scientists, kept mind out of all other realms, including biological processes such as procreation, self-healing properties, and the structure of DNA. Thoughts and ideas dwelt in a realm different and separate from the physical world.

The anthropic principle, however, suggests that there is a link between the overall design of the universe and the human mind, which can *recognize* the design and, further, exists because of it. According to this principle—which links mathematical fundamental constants to both quantum physics and cosmology—there is an underlying causal principle, what Aristotle called τειoδ (*telos*), purposeful action, animating the universe. Amit Goswami (1993) postulates, "The universe becomes self-aware through us."[4] This idea echoes that of Gurdjieff, who suggests that humans have come into being to serve a higher purpose, namely to help the Earth, solar system, and galaxy evolve. He links this to an idea he calls "reciprocal maintenance." Gurdjieff biographer J. G. Bennett further explains Gurdjieff's contention that, like every other organism, "[m]an is an apparatus for 'the transformation of energy' and he is specifically required to produce sensitive and conscious energy for maintaining the harmony of the solar system."[5]

COINCIDENCE OF MATTER AND MIND

Let us assume that the label Pythagorean/Platonic implies a rejection or denial of "matter" as ultimate reality, and [is] instead an affirmation of a search for some grand design . . . [for] which our vocabulary might not be quite adequate. We might call it "mind" or "spirit," or "supermind," or whatever other word might appear appropriate when we attempt to speak about the ultimate roots of reality. The fact that the New Physics appears to point in this direction has been repeatedly stressed by Werner Heisenberg, who sees a parallel in the implications of quantum physics and the shift from Democrit's extreme materialism to the . . . emphasis on mathematical form, and to link mind or consciousness to important aspects of reality.

E. Gora (1983)

The idea of the importance of coincidences, as such, was introduced by Paul Kammerer in 1920, in his book *Seriality,* in which he logged a hundred amazing examples. His complex idea intrigued Einstein and was expanded by Carl Jung, who changed Kammerer's term to the more widely used word *synchronicity,* or "meaningful coincidence." Like Kammerer, Jung noticed that if two events were not causally related, but connected by *meaning,* it therefore established that a human mind was required to see the connection.

In physics there are key numerical coincidences connecting the microcosmic world to the macrocosmic world, a mathematical link between certain ratios in both atomic and galactic structures. The brilliant physicist Edwin Gora—who wrote two watershed articles for *MetaScience Quarterly* on the subject of Pythagorean trends in modern physics—paired the gnostic concept of "Aeons," emanations from the first cause, or the power of the Absolute, to "the symmetry principles of the New Physics"[6] and the breaking of that symmetry with the process of creation.

Gora's doctoral mentor, Heisenberg,[7] tells us that Sommerfeld "believes in numerical links, almost in a kind of number mysticism of the kind that Pythagoreans applied to the harmonies of vibrating strings. That's why many of us have called this side of his science 'ato-mysticism' though, as far as I can tell, no one has been able to suggest anything better."[8]

There is no causal reason for relationships between certain ratios in atomic and galactic structures except for the fact that humans notice the coincidental link. For instance, Sommerfeld's number 137, found in the fine structure constant, which measures the ratio of matter to energy, as well as the strength of electromagnetic force inside of atoms (1/137 is the probability that an electron will absorb a photon), shows up also in the spin of the electron and in the expansion rate of the universe. There is no known intrinsic reason why these separate realms would use the same number. It was for this reason that 137 also fascinated other physicists, such as Wolfgang Pauli and Richard Feynman. If the situations are in fact related, this would suggest an overarching design pattern to the structure of the cosmos whereby subatomic meets macrocosmic.

THE ORIGIN OF LIFE

The human mind did not originate with the beginning of biological time. Qualities of mind must have existed before that or life itself could not have evolved. Some theorize that had the universe unfolded in even a slightly different fashion, the human mind would not have evolved. Richard Morris, in *The Edges of Science* (1990, p. 213), notes that had there not been "an unstable form of beryllium," this element would not have combined with helium to produce high levels of carbon, and without carbon, great amounts of oxygen, and ultimately organic molecules, would also not have formed. And even before such a development as a complex molecule, it would take great star systems billions of years to begin to generate the other major elements required for life, several billion years more for planets to cool, to make them hospitable for life, and then, of course, there would have to be water and lightning storms and the distance from the sun would also have to be just right. For the authors of *The Anthropic Cosmological Principle*—astronomer John Barrow and mathematician Frank Tippler (along with the writer of their foreword, theoretical physicist John Wheeler)—requiring that this simply be the end product of coincidence is asking for too much: "That is the central point," namely that "a life-giving factor lies at the center of the whole machinery and design of the world."[9]

In the beginning there was, most likely, an incredible explosion of a unified mass, and the universe was born. Clusters of matter bound by gravity interwove and formed galaxies. These galaxies gave birth to smaller sub-wholes known as solar systems, our particular arrangement consisting of nine planets, many of which have one or more moons, circling the Sun in roughly the same plane.

The entire universe is a hierarchical structure that is always in motion. Subunits are delineated by their various levels of organization. One particular level gave birth to life here on Earth. As part of the hierarchy, the Earth is composed of smaller units called atoms. Made up of elementary particles (the electron, proton, neutron, and their subatomic precursors), the various arrangements of these atoms form the elements, the building blocks of matter and life. The structure, composition, and position of the Earth were prerequisites for life. Each of these coordinates is just as important as the others in understanding biological emergence.

As opposed to Cartesian dualism, the theory of the anthropic principle suggests that volitional activity inherent in the structure of matter eventually evolved into amino acids, DNA, and one-celled organisms. Consciousness evolved with the increase of volitional developments in the nervous system. This occurred in the paramecium when it moved toward the Sun to obtain warmth; in insects, fish, reptiles, and mammals when adaptive instincts emerged; and in humans when they became tool-makers and developed language. That we are a product of the universe suggests that the intentional aspects of our components, most notably the purposeful interaction of elementary particles, gave rise to the self-direction inherent in DNA and the zygote. Eventually these lawful processes evolved into self-awareness and consciousness as we know them.

DEVELOPING A THEORY OF CONSCIOUSNESS

A comprehensive theory of consciousness should lay a foundation for coming to terms with the following questions:

1. How can the psyche infuse itself into the brain?
2. What is the relationship of consciousness to the four physical forces of the universe: gravity, electromagnetism, and the strong and weak nuclear forces?
3. Can consciousness be considered a fifth force, or is it an outgrowth of the other four?
4. Can matter by definition be conscious, or can components of consciousness be inherent in matter?

In order to begin to tackle these questions, it is best that we define the term *consciousness*. In looking through various dictionaries, reading the ideas of others, and discussing the word in a number of college classes, I have come to the realization that the act of becoming conscious is a complex process that has many attributes. Clearly, "that luminescent presence of coming-into-being"[10] involves a whole host of variables (see figure 1.1), the ultimate one perhaps being the act of self-awareness. We can state that humans are the most conscious animals because our powers of self-perception, thought, verbalization, intention, and so on are more highly developed than in other forms of life.

CONSCIOUSNESS

A complex term encompassing the following:

Awake

Awareness . . . of an external event or internal physical or psychological state (Descartes)

Sensitivity, knowing, perceiving, apprehending, remembering

Involving rational abilities

Conscious of being conscious (Lachman)

The ability to think in words: that is, in a complex symbolic form that can be communicated to other minds

Mind in the broadest sense

A unitive process encompassing self-reflective or self-referential and transcendent functions (Goswami)

The totality in psychology of sensations, perceptions, ideas, attitudes, and feelings

Conscious, preconscious, unconscious, and collective unconscious states, each with its own "consciousness" (Freud, Jung)[11]

Capable of:

Decision making	Design
Ideation	Thought
Organization	Communication
Perception	Discrimination
Reflection	Volition
Sensation	Self-observation
Planning	Emotion
Negentropy	Sympathy
Empathy	Will
Purpose	Teleology
Entelechy	

Fig. 1.1. The many components of consciousness

Consciousness is not an either/or concept. The act of becoming conscious lies on a continuum, starting with simple awareness and ending with advanced thinking and volitional activity. Even the first one-celled organism that moved itself into the warmth of the Sun was to some degree conscious. Certainly perception, purpose, awareness, and decision making were evident, even if the one-celled being reacted "automatically" or instinctively. Something inside that organism was conscious (or programmed by conscious forces) to some extent. This "something," which Freud would call the unconscious, "thinks."

Herbert Read and Jean Piaget hypothesize that humans evolved from lower animals because of intentional movements.

> Man has not reached his present superior status in the evolution of the species by force alone, or even by adjustment to changes in the environment. He has reached it by the development of conscious-ness, thus enabling him to discriminate the quality of things.

Read goes on to state that Piaget links intelligence to the organ-ism's initial reaction to the environment:

> Intention is the essential characteristic of intelligence. . . . Piaget shows that intentional adaptation begins as soon as the child tran-scends the level of simple corporal activities such as sucking itself, listening, looking and grasping, and acts upon things and uses the interrelationships of things.[12]

The neurophysiologist A. R. Luria links consciousness and inten-tional adaptation to the onset of language and the ability to think in words. Once the left temporal lobe adapted itself to specialize in language—and there is great debate as to when this occurred—humans, free from the present, were able to represent both the outside world and interior states in mental symbolic fashion. This enabled them to begin to *manipulate concepts* instead of actual physical things. Memory was further enhanced and rational thought was able to advance at a more rapid rate. The new generation was able to stand on the shoulders of its ancestors. While all other animals are bound by instinctual forces and the immediate present, humans are able to reflect on the past, consider multifaceted aspects of the present, and project into any of a variety of possible futures. This increase in linguistic ability caused a corresponding increase in cerebral complexity. With the advent of writing, movable type, mass communication, and now the Internet, this process has continued its evolution at an ever-increasing rate.

Mind and Matter

Insofar as the mind can know matter, it has a group structure isomorphic to that of matter.

ARTHUR YOUNG ON ARTHUR EDDINGTON
IN MISHLOVE'S *ROOTS OF CONSCIOUSNESS* (ON LINE)

One of the most important thinkers involved in the discussion of the link between mind and matter is the French Jewish philosopher Henri Bergson (1859–1941), Nobel Prize winner in literature. Son of a Polish musician on his father's side with a mother from Ireland, Bergson's great contribution was his idea of the *élan vital,* which is a creative life force imbued in matter that propels matter to, in a sense, escape its own confines by expressing an impulse, as expressed in humans, to move toward "novelty, freedom, and self-direction." Author Gary Lachman notes in his book *A Secret History of Consciousness* that Bergson's "defining characteristic of mind is that 'it has the faculty of drawing from itself more than it contains.' This, for the materialist-mechanistic science that wishes to explain consciousness is an impossibility." Where matter "restricts life's impulses and scatters its energies," the élan vitale propels life to generate food in the case of plants through photosynthesis and motivates humans to escape matter through creative endeavors, yet at the same time imbue more of consciousness in matter through the human expression.[13] We see here Bergson producing a profound idea that somehow consciousness as a force creates something beyond what has already existed, and at the same time, we see evidence of Bergson's idea that humans as an instrument of consciousness begin to imbue more consciousness back into matter as seen in such inventions as the radio, television, computers, nanotechnology, and artificial intelligence.

This idea was realized by Nikola Tesla, who not only invented the first remote-controlled robot but also envisioned this invention as "a new species on the planet," not made of "flesh and bones" but rather of "wires and steel."[14] Here, in 1898, was a culmination of Bergson's vision.

Stepping back into the realm of biophysics, we can state with certainty that DNA's ability to direct the metabolism of the cell, produce the proper enzymes and amino acids, replicate itself, and also ultimately orchestrate the development of the fertilized egg into a fully developed organism is a conscious display of the highest order. Memory, intent, organization, awareness, design, and purpose are each fully developed in this instance. The motive force inherent in DNA is a form of intelligence and its structure is imbued with consciousness. Although the nature of its consciousness is in many qualitative ways different from the psyche of our brains, it is DNA that directs the development of the human psyche. Thus, it may be considered a more primary form of consciousness.

The MIT professor of computer sciences and self-made millionaire Ed Fredkin views DNA as "a good example of digitally encoded information" or, as R. S. Jones (1982) puts it, "consciousness reflected in matter."[15] It is Fredkin's hypothesis that information is even more primary than matter and energy. Subatomic particles, according to this view, can be seen as "bits of information," just like those found inside "a personal computer or pocket calculator. . . . The behavior of those bits, and thus, the entire universe," Fredkin says, "[is] governed by a single programming rule."[16] Through eternal recapitulation and incremental transformations, the "pervasive complexity" that we see as life emerges.

> The more I examine it and study the details of its architecture, the more evidence I find that the universe in some sense must have known that we were coming.[17]

David Chalmers (professor of philosophy and director of the Centre for Consciousness at the Australian National University), in a seminal article on consciousness in *Scientific American* in 1995, echoes this idea by stating that "the laws of physics might ultimately be cast in informational terms. . . . It may even be that a theory of physics and a theory of consciousness could eventually be consolidated into a single grander theory of information."[18] This idea had already been expanded by biochemist, philosopher, and cancer researcher Alfred Taylor, who served as the head of cancer research at the Biochemical Institute at the University of Texas from 1940 to 1965. He writes on science and philosophy in his article "Meaning and Matter." It is Taylor's supposition that since all matter is derived from "a common source . . . we are forced to the conclusion that *organization* is the determining factor, whether energy appears as hydrogen, lead, a daisy, or a man. Something must distinguish one from the other, and that something is organization, meaning consciousness."[19] For Taylor, life quite simply cannot be a chance process.

In the early 1950s, Crick and Watson uncovered the basic structure of DNA. They discovered the double helix, a spiral-structured tetragrammatic molecule containing phosphorus, oxygen, carbon, hydrogen, and nitrogen. The particular arrangement of its four molecular bases—thiamine, adenine, guanine, and cytosine (TAGC)—codes for every plant and animal from virus to human. It is simply the sequencing of the two base pairs, AT or TA and GC or CG, along the sugar phosphate backbone that contains the program for the construction of the particular life-form in question. In the complex binary program, in which TA is one unit and GC is the other, the only difference between the zygote of a human and that of a pterodactyl is the base sequence of these four molecules!

The discovery of the structure of DNA has brought the concept of consciousness down to the level of the atom, for it is the particular arrangement of specific atoms that codes for, and thereby directs, all forms of life.

Further proof that atoms are involved in processing conscious information can be found in the field of neurophysiology. For instance, Holger Hyden (1964) has discovered that when learning takes place, messenger RNA (mRNA) changes its base count in neural and glial cells of the brain. Messenger RNA is the liaison between the DNA molecules inside the nucleus of a cell and the various components within the cell body. After a period of time, Hyden discovered, the mRNA directs the production of protein chains on the ends of the dendrites extending from the neurons to house the new encoded memory trace.[20]

Other neurological research establishes that different types of cognitive processes are encoded in, or triggered by, specialized molecules called neurotransmitters. For instance, when an animal is in a fearful situation, adrenaline is pumped into the brain. When a human dreams, serotonin and melatonin, produced from the pineal gland, are involved. Just as with the basic components of DNA, the neurotransmitters are made from just a few basic elements: oxygen (O), hydrogen (H), nitrogen (N), and carbon (C). Serotonin and melatonin have a molecular structure known as the indole ring—a combination of a carbon ring and a pentagon-shaped ring with nitrogen at its nadir—which is also found in the psychotropic drug LSD. Fifty micrograms (fifty millionths of a gram) of LSD is enough to alter one's consciousness in dramatic and awe-inspiring ways. Our state of consciousness is based upon a fragile neurochemical equilibrium.

Our search for our "mind" has taken us to the basic structure of

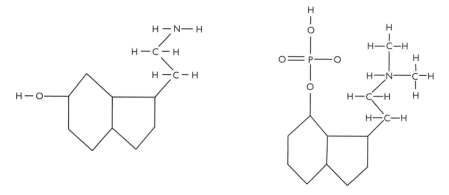

Fig. 1.2. Serotonin (left) and LSD (right) both have the indole ring.

matter. Going back to our definition of consciousness, we can see that elementary particles, atoms, elements, and molecules contain components of consciousness, not only because they can be utilized to program cognitive processes, but also because they house within their structure the capacity for the following attributes:

> Basic awareness
> Organization
> Lawful design
> Discrimination

And even, perhaps, intention, purpose, memory, and communication.

Taylor points out that the bodies of living organisms are constantly turning themselves over. In a human being, the components of every cell change every seven years.

> Since the matter aspect of the body is constantly changing, this fact alone discredits the idea that matter is the primary value. . . . How then can consciousness or intelligence be a mere product of the functioning of the nervous system when this system is compounded of transitory materials? The *meaning* of the form transcends matter changes. The same being continues but not the same materials. . . .

> The universe is an organized system. . . . The principle of progressive increase in [order and] meaning is evident in both organic and mineral evolution.[21]

Surely the interaction of electrons with photons, protons, and neutrons is a highly ordered procedure. Somehow, within the structure of the electron it "knows" that it must repel other electrons and be attracted to positively charged protons. Decision making occurs at the level of the electron whether or not the electron itself "thinks." One way or another, it is programmed to respond in a predictable and lawful way. There is a basic awareness inherent within the construction of the electron, for if this were not true, the structure of matter would have no order. The very fact that the periodic table of elements exists is proof of conscious design, purpose, order, and intent in the creation of the elements. By definition, since we see components of consciousness within the structure of matter, we can therefore conclude that aspects of consciousness are inherent there as well.

Cosmic Law and the Universal Forces

Our entire quest for scientific truths is based upon the tacit assumption that the universe operates lawfully. *Lawful interactions presuppose conscious design.* The very fact that the planets circle the Sun in prescribed paths equivalent to Kepler's harmonic law P^2/D^3 (where P = period to circle Sun and D = distance between planet and Sun) is proof of conscious design within the structure of the universe. Neatly sidestepped by neo-Darwinian paradigms, which suggest that the emergence of life is a chance process, this self-evident truth was known by most, if not all, of the great scientists of the past.

Attributes of consciousness are evident not only in lower forms of life, but also on cosmological levels. Physics has uncovered four forces of the universe:

1. Electromagnetism: force that holds molecules together, the sharing of photons by elementary particles.
2. Gravity: force that holds the planets together.
3. Strong nuclear force: force that holds the nucleus together.
4. Weak nuclear force: force that holds the neutron together.

It is stated by the physicists that all known physical properties can be reduced to these four forces. Arthur Young speculates that all four may ultimately derive from the spin and other properties of the photon. Be that as it may, there is one basic component of the universe that is not included in these forces, and that is the motive power behind it, animating it.

Consciousness as a fifth force may be looked at as the backdrop of an intentional lawful cosmic mosaic that corresponds to the élan vital of Henri Bergson; it is the motive force that drives the universe. It is also the purposeful or thoughtful power that gives rise not only to spacetime and the four physical forces, but also to the emergence of life.

From the discussion above, it is clear that consciousness as a force or attribute of the cosmos did not suddenly appear with the dawn of human beings, nor did it begin with the first one-celled organism. It was there from the start. Not only are biological organisms "intelligent," but the motion of the planets and the very structure of matter are also evidence for psychical design. In that sense we can see that evolution is also a form of devolution, as the highest principles of consciousness must have been present from the start.

From this point of view, equations and inventions are not so much created by humans as they are discovered by them. The airplane and flight to the moon happened because human beings looked out at the world and saw that other animals could fly. A human could not run a four-minute mile or design a computer unless the respective abilities were already inherent as distinct possibilities from the outset.

HIERARCHY OF MIND

Having established a relationship between the human mind and the structure of matter, we can now turn our attention to the qualitative differences between the inherent attributes of consciousness in humans and atoms. Any attempt at modeling the psyche must certainly address itself to the question of the various levels of mind. The human brain can be separated into three basic levels:

1. The Physical Level. The realm of physics and the four known forces of the universe; the physical atoms (and subatomic and elementary particles) that make up the brain.

2. The Biological Level. Biophysics, the realm of life; the development of amino acids, DNA, the structure of the cell, and so on, including the development of neurotransmitters and a neuronal network in higher animals. We could also include here the primary instincts and automatisms.

3. The Psychospiritual Level. The realm of psychology as defined by such writers as Pavlov, Skinner, Freud, and Jung; the area of higher states of consciousness, such as will, psychology, as delineated by Ouspensky and Gurdjieff, and more esoteric realms, such as the development of clairvoyant powers, interaction with one's soul, and self-transcendence as espoused by the Austrian philosopher and metaphysician Rudolf Steiner; realms discussed by religious doctrine.

If we take into account evolution, the expansion rate of the universe, and chronology, we could add a fourth component:

4. The Human Being's Hierarchical Time and Place in the Cosmos. Specific factors such as the position, structure, temperature, and other attributes of the Earth; the particular teleological chain of events that led to the development of life; and the emergence of humans at this point in time. The animating

principle of the universe, that is, the first cause, falls into this category, as does our place in the intelligence hierarchy.

These levels of mind can be arranged hierarchically, and each has its own organizing principles. The relationships between levels are antisymmetric. Moving down the hierarchy, components become more detailed and specific (for example, the chemical structure of neurotransmitters is a primitive form of consciousness, but more advanced than the processes involved with the interaction of elementary particles). Movement up the hierarchy is toward greater holism, such as from atoms to molecules, to DNA, to a brain, with the tipping point being self-awareness. Threshold values separate one realm from another. In the case of the three basic levels mentioned above—physical, biological, and psychospiritual—bioelectric forces seem to be specifically utilized as a medium of communication between each stratum. When thoughts become physical actions, the transfer of electrons (during a nervous impulse) carries the message from the mind (software) to the brain (hardware) and then to the body. Interestingly, the psychological concept known as the *will* can be seen as a liaison between the mental and physical domains.[22]

I. Physical Level

At the level of the atom, there is very little, if any "free will," although there is some measure of randomness or indeterminacy. This realm involves the laws of physics, as well as the most fundamental feature of consciousness: basic awareness, or sensitivity. Other primary components of consciousness operating at this level are discrimination, organization, and intentionality of some sort. For example, opposite poles of magnets attract each other and same poles repel. Electrons are attracted to protons but are repelled by other electrons. Magnets create highly ordered fields, and elementary particles aggregate into the highly ordered periodic table of elements.

II. Biological Level

This realm involves goal-directed (negentropic) behavior and thus a more complex "conscious" mechanism. The processes of growth, propagation, evolution, and intention exist at the level of DNA. In fact, all life processes can be seen as teleological, since they purposefully take into account future needs.

Although life functions are quite different from the relatively simple interaction of elementary particles, or the aggregation of atoms into the elements and complex molecules, there are also basic similarities between, for instance, the laws of chemistry and photosynthesis. The level of biophysics, although more sophisticated than simple chemical interactions, is similar in that prescribed patterns of atomic interactions follow lawful procedures. The inventor Nikola Tesla (1856–1943) pointed out that the growth of crystals contains within it precursors for a life-forming principle. The major difference between Level I and Level II is one of increased volitional ability and the accompanying so-called spark of life, the élan vital of Bergson. In the case of the appearance of plants, the life-giving process directly stems from the transformation of solar molecules (photons), in combination with water and earth, into organized cells that can be eaten by other organisms. In this sense, all of life is made up of bits of the sun combined with particles from the Earth. Looked at from a strictly atomic point of view, DNA, the building block of life, is made from five elements (carbon, oxygen, hydrogen, nitrogen, and phosphorus) and light. DNA has more of its own say in its destiny than do the elementary particles, yet DNA is a hierarchical construction of these particles.

There is a qualitative shift or quantum leap occurring between nonbiological and biological molecules, inorganic and organic. The difference involves a number of variables including a greater use of the element carbon; use of water, sunlight, and electricity; some form of sense perception and a memory system of sorts; unitive field properties for processing data; a mechanism for ingestion of organic material and elimination of waste products; some form of locomotion and growth; and a push for self-preservation and procreation.

The big difference between plants and animals is an increased level of autonomy. One could make the case that the heliotropic aspect, that is, the ability of leaves to turn toward the Sun, evolved into muscles. Where plants are rooted to the earth, the first organisms, such as the paramecium, already had some more advanced level of autonomy because of their ability to swim in the primordial soup.

III. Psychospiritual Level

The cyberspace of the psyche as delineated by such mind psychologists as Freud and Jung describes processes of our existence that seem to bear little connection to the so-called physical world. Nevertheless, Jung states that by its nature, the self arises and connects the inner mental realm to that of the outer physical. Behavior psychologists such as Pavlov and Skinner would argue that internal processes such as thinking and dreaming are based upon reflex action and associative mechanisms only, but this view essentially ignores the mind and the unconscious and focuses on manifest behavior.

In a sense, Gurdjieff and Ouspensky combine mind and behavior psychologies. Concerning behaviorism, they essentially agree with Pavlov and Skinner that much of human thought processes is mechanical. Most of our actions are due to automatic responses to stimuli. Gurdjieff and Ouspensky write that we spend most, if not all, of our life in an automatic pilot existence, which they call "waking-sleep." This mechanical state is quite similar to the behavior of inorganic or organic chemical reactions in that no real "thought" is claimed to be involved. The human/machine simply moves in a prescribed stimulus-response path.

"But there is a possibility of ceasing to be a machine," Gurdjieff says. "It is this we must think of and not about the different kinds of machines that exist."[23] Gurdjieff tells us that highest states of consciousness are equated with self-evolution, transformation, and acts of

Fig. 1.3. Jesus in the Temple, *Heinz Hoffmann, 1850*

one's own willpower. The more a person directs his or her fate, the higher the state of consciousness.

I AM

The highest states of consciousness involve greater autonomy and creative living, what Maslow calls "self-actualization" and accompanying peak experiences. These transcendent feelings encompass a timeless sense of oneness with the universe, such as you might experience on a warm crystal-clear night while lying on the ground and staring up at the stars.

One of the most difficult problems in explaining the ultimate mystery of consciousness, specifically human consciousness, your consciousness and mine—which Chalmers calls "the hard problem"—is how the neuroelectric processes of the brain create the *subjective* sense of "I."[24] The "easy problem," which is not so easy, is what Crick and Koch call the "unity of consciousness."[25] For instance, as I type this very passage, the radio is playing *Crazy For You,* my wife has put me in charge of the oven and I can smell the banana bread that she is baking, a somewhat cloudy day is apparent outside the window, it's July 4, 2007, and I'm thinking about the upcoming barbecue we are going to and the fireworks tonight and also considering how to integrate the easy and hard problem into this section of the book. All of these factors and different modalities are unified in my mind—and how the brain does this is the easy problem! Neuroscientist and Nobel Prize winner Eric Kandel sees the solution to the easy problem as a neurological binding issue: how various neural networks combine to create a single viewpoint.

As for the hard problem, while some researchers think that the sense of "I" is located in the thalamus, the main "switchboard" in the brain, Crick suggests looking at the claustrum, a part of the brain that evolved or split off from the amygdala, the seat of temperament, personality, and certain forms of aggression, such as rage. The claustrum, working in concert with the amygdala and insula, helps unify conscious and unconscious/emotional experiences including life-threatening ones. Liotti et al. suggest that this is how consciousness itself would have evolved, that is, in reaction to such things as fight-or-flight situations.[26] Ultimately, all of these areas would still have to be modified by higher centers in the frontal cortex and the thalamus, or central processor, so Crick's theory doesn't really usurp the thalamus as the true neurological/visceral source of the sense of "I."

The Sufi, Gurdjieff, quantum physicist David Bohm, and Rudolf Steiner have an answer to the hard-problem conundrum, that is, the problem of understanding the subjective sense of self. The key, which stems from 600 BC to the teachings of Zoroaster, is found in the Hebrew scriptures where Moses confronts the burning bush. God tells Moses his name: "I AM THAT I AM." There is only one "I AM," one God, one universe. All are extensions of that One, which is expressed by Leibniz's monad theory, in which the microcosm reflects the macrocosm, or by the holographic universe, in which each part codes for the whole. Steiner explains that a person cannot say "I" for another.[27] For each of us, our own expression of selfhood, the simple proclamation that "I am," that "I" exist—which we can each make only for ourselves—asserts our connection to the great "I AM" from which we all spring. The path to higher consciousness can be found only through the self—and thus, the connection to the One.

One touch of nature makes the whole world kin.

JOHN MUIR

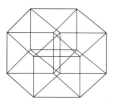

2

PARAPSYCHOLOGY
AND ESOTERIC THOUGHT

This chapter originally appeared in *Parapsychology Review* in May–June 1981. It has been left essentially unchanged. A few thoughts come to mind regarding its content, which is a review of books on parapsychology and esoteric thought written between the years 1873 and 1925. First, the field known as parapsychology has all but disappeared, only to reemerge under the less threatening banner of "consciousness studies." On the positive side, participants in this new field have produced many successful conferences. On the negative side, few, if any, university professors are allowed to study such things as telepathy, psychokinesis, and precognition without risking their careers. Almost every major textbook of introductory psychology mentions ESP (extrasensory perception), but none, to my knowledge, takes it seriously. Further, today's mainstream bookstores, although much larger than they were twenty-five years ago, carry woefully inadequate resources for obtaining any sober treatise on scientific aspects of psychic research.

This chapter is dedicated to my spiritual compadre, Robert Adsit, a gifted artist and extraordinary individual who discovered many of the books I discuss. Robert lived for many years in a dingy storefront on East 9th Street on the Lower East Side of New York City. Robert's flat was a treasure house of occult books and artifacts, his own complex abstract landscape etchings, watercolors, and oil paintings, a piano that he occasionally played, various and sundry street art, and also a lot of pure junk. His greatest find was a framed signature, circa 1945, by Salvador Dali.

One of the most intriguing tomes that Robert had found was called *Psychical Developments,* by E. H. Anderson, written in 1901. This book was too important for me to keep on loan for too long, so I returned it sometime around 1980, only to hear that it was later stolen from Robert, along with an old-time radio and some other artifacts, including the great Dali signature.

As friendships sometimes dwindle, I lost touch with Robert about twenty years after we had first met. Later, after trying to contact him repeatedly to no avail, I found that he had passed away, although he was only fifty-four. A few weeks later, I was surprised to receive a pack-

Fig. 2.1. Robert Adsit, artist extraordinaire, circa 1973

age from his roommate. It contained one of Robert's most eloquent watercolors neatly framed. I was staggered by its subtle power when I opened the box in the post office.

Just a few months ago, when I was scouring the websites of secondhand bookstores on the Internet, I came upon an available copy of E. H. Anderson's small masterwork, and purchased it immediately for easily twenty times what Robert had paid for it way back when.

Next to Robert's shop was a Tibetan store, which contained statues and other genuine artifacts from that exotic land. There, in 1973, I discovered, for fifty cents, a strange paperback published in 1957 called *The Third Eye,* the autobiography of the Tibetan lama Lobsang Rampa. The cover depicted a typical 1950s photograph of a Western man who resembled Orson Welles, with a glass eye air-brushed onto the center of his forehead.

I kept the book unread for about a year while I worked on my master's thesis, at the University of Chicago, entitled *Levels of Mind.* I had started with a neurological study, the difference between left and right hemispheres, holographic brain theory, and the link between brainwaves and states of consciousness. Then I moved on to Freud's study of the unconscious and the model of the psyche outlined in my book *Inward Journey,* followed by Jung's theory on the archetypes and collective unconscious; J. B. Rhine's scientific studies in psychokinesis and thought transference; F. W. H. Myers's work in life after death and his theory of the universal mind; a discussion of the link between astrology and neurophysiology; and some of the work discussed below. Finally, after finishing the hundred-page thesis, I opened *The Third Eye,* and could not put it down. It is a fabulous story about a young psychic Tibetan boy growing up in the 1930s, who traveled to Lhasa to become an aide to the Dalai Lama because of his ability to read auras. But who was that white guy on the cover?

Little did I know that not only was Rampa the author of fifteen other books on his life and Tibetan metaphysics, but also, as I would come to learn in his third book, *The Rampa Story* (1960), the author really was the Caucasian man on the cover of the first book. His name was Cyril Hoskins. As the story goes, Hoskins was a British plumber who was disheartened with life and was suicidal. The real Rampa, the Tibetan, was very ill and near death. The Great Masters made a deal with Hoskins. They promised to take him to heaven in exchange for his body, and he agreed. Now Rampa could stay on to complete his work, which was to enlighten the Western world as to the nature of our higher abilities.

The story, of course, was so astonishing that *The Third Eye* was branded a hoax, even though it had sold 150,000 copies on its first hardcover run in 1956.

Having read hundreds of treatises on higher states of consciousness, both Eastern and Western, I will boldly assert that the Rampa books are unparalleled in their discussion of esoteric truths. At the same time, they present a truly enchanting account of a fascinating young boy and how he grew up to assist the Dalai Lama (the one before the present Dalai Lama), he worked in China as a doctor, and then eventually escaped to the West. Whether Hoskins was really taken over by a master Tibetan, I cannot say. My scientific side remains skeptical.

Like so many other intriguing and thought-provoking metaphysical works, the Rampa books cannot be found in any modern mainstream bookstore. The new generation is being robbed because it cannot easily find the classic works that I had such ready access to a quarter century ago, books by such authors as Madame Blavatsky, Alice Bailey, Jane Roberts, Wilhelm Reich, Marc Edmund Jones, Dane Rudhyar, Lyall Watson, Charles Panati, Ostrander and Schroeder, Gurdjieff and Ouspensky. True, some of these books are still in print somewhere, and all, or almost all, are available in secondhand bookstores online. But how would neophytes know to look for such classics if they had no way to evaluate them by holding them in their hands? Surely these works are legions better than the junk that invades the space in most New Age and metaphysical racks in the modern bookstores of today.

Fig. 2.2. Mountain Home *by Robert Adsit, 1976*

TURN-OF-THE-CENTURY THEORIES ON LAWS OF MIND

Man can attain any position in life he may desire, if he only believes that he can attain it, and has the courage to follow that desire. He can make his way into the most exclusive circles, he can surround himself with men of great ability, he can obtain all of the good things of this world that he desires. These things are possible simply because man is all mind and the power of mind is unlimited. . . . Remember, man is at one with the universal; his natural condition would be that of harmony with all.

E. H. ANDERSON,
A MENTAL SCIENTIST, 1901

Parapsychologists are seekers of truth. Many in this group were skeptics who turned to the field either to disprove the various claims of psychics or out of some innate curiosity, triggered perhaps by an anecdote, psychedelic experience, or mystical legend from the ancients. After the initial surprise of discovering, for example, that telepathy actually is a human ability, the seeker is trapped. All of a sudden he or she is thrown into a world of mind reading and psychic metal-bending, out-of-body experiences, life-after-death studies, and even extraterrestrial communication. This author has exclaimed many times that "all I wanted to prove was that humans were telepathic" . . . but Pandora's box lies waiting. It opens randomly with spontaneous eruptions of paranormalacy and—as Stanley Krippner notes (in his book of the same title)—the "song of the siren" has begun. Homer's *Odyssey,* echoed perhaps by modern writers such as Martin Ebon, warns modern explorers to steer clear of the sirens, for there are hidden reefs in this uncharted territory. This separate reality may have its own laws different from what we call the everyday world. Nevertheless, a question arises as to whether or not this dimension really is uncharted, that is: Are there people in history who understood the laws of mind and can shed light for us twenty-first-century initiates?

The potential for answers to the "siren's song" is the impetus that leads many of us beyond Freud and Jung and into the dusty metaphysical stacks in well-known or unknown libraries and the parapsychology sections of secondhand bookstores, for it is there that the Susy Smiths, Jane Robertses, and J. B. Rhines, and the Arthur Koestlers of the past still lurk. It is there that the ancient wisdom is carefully preserved, only to be rediscovered again and again.

Surprisingly, researchers from nearly a century ago, and even before 1900, also worked with what we consider "modern" techniques, such as psychotronic devices and so-called Kirlian photography. The famous palmist Cheiro (Count Hamon), in an 1898 version of his classic work *Language of the Hand,* described a scientist from France who had developed a small device that moved when one willed it to do so.

Unlike Cheiro, however, most of the authors cited below are virtually unknown today. Various obscure texts preserve the metaphysical secrets that are so often ignored or glossed over today.

In 1911, in a book entitled *Photographing the Invisible,* James Coates, a member of the American Society for Psychic Research, describes the precursor to Seymon Kirlian, a Mr. Kilner, who photographed "N-rays" issuing from his body. These "N-rays," like the Kirlian aura, purportedly showed the emotional attitude and state of health of the subject.[1]

Thought Transference

> *The most perfect condition either for the conveyance or the reception of telepathic impressions or communications is that of natural sleep.*
>
> The Law of Psychic Phenomena:
> A Working Hypothesis,
> Thomson Jay Hudson, 1895

Eli Beers, another obscure psychic researcher, from Chicago, whose coverless paperback was given to me by a garage sale addict, writes in his 1914 work, *Mind as a Cause and Cure of Disease,* about studies in dermovision: Dr. William Carpenter, a distinguished writer on physiology, speaks of people who, although blind, could "distinguish colors of surfaces, which were similar in other respects, and says that they were probably able to do this because their sense of touch was so acute as to enable them to distinguish . . . the absorption of some rays and the reflection of others."[2] He also discusses telepathy, psychic healing, and metaphysics. Concerning thought transference, Beers suggests that telepathy may in fact be the only way that one person transfers a concept to another: "Nor, as far as we know, is there any reason to suppose that thought itself can actually be transferred from one individual to another by any other mode of communication between man and man. That is to say, thought cannot be transferred from one person to another in the same manner as a material substance can, where the identical thing, as for example a book, passes one to the other."[3]

Certainly we do not really understand how cognition occurs, even though we can trace its neurophysical path from protein chains on up to observed behavior.[4] Beers writes that there is a difference between transferring physical objects and transferring spoken mental concepts. The latter are "sent" from one person to another "through the medium of sound and by use of symbols . . . Furthermore, since these symbols could be translated from one language to another—that is, from one form of symbology into another . . . and so handed down from generation to generation . . . thus man by his inventions has broken down the barriers of space and time."[5]

This idea concerning the primacy of symbols as the mechanism for communication via a telepathic process is comparable to Jung's ideas of the autonomous complex.

> We do not feel as if we were producing the dreams, but rather as if they came to us. They do not submit to our direction but obey their own laws. Obviously, they are autonomous complexes, which form themselves by their own methods. Their motivation is unconscious. We must say therefore that they come from the unconscious. Thus, we must admit the existence of independent psychic complexes, escaping the control of our consciousness and appearing and disappearing according to their own laws.[6]

Jung goes on to say that when these autonomous complexes appear as "supernatural," they derive from the collective unconscious. "Being strange to the ego, they always appear as if externalized."[7] Although he would later modify his thinking, Jung, at the time of writing, was not linking these complexes to paranormal communication. Beers, however, was quite open to the idea of thought transference. For instance, he describes incidental cases of telepathy, such as thinking of someone a minute before he appears, or staring at someone in a crowd only to have him somehow sense this and therefore turn to look back at you. In discussing the work of the London Society for Psychical Research, Beers concludes that one mind can directly influence the mind of another.

Just as identical tuning forks or pendulums vibrate in synchrony with one another, telepathic exchanges between two minds may involve similar principles. Beers states, "[A]lthough the ability of a magnet to cause iron filings to become magnetized can be seen, in telepathy the medium of communication is unknown." He then theorizes that "for every thought there is a corresponding motion of the particles of the brain."[8] Therefore, Beers's resonance theory provides some physical basis for a medium of telepathic transfer. Like Thomson Jay Hudson, quoted above, Beers would also use this simple idea of resonance to explain how the power of healing could be generated in another as well.

The Universal Center

A small windfall by Kate Boheme, entitled *Realization Made Easy* (1917), was discovered at a rummage sale in Cambridge, Massachusetts. Boheme was a leader of a psychic society from Holyoke, Massachusetts. She tells us that the feeling of *power* is dependent upon the manner of *thinking* of energy. When someone feels his energy, there is a realization that the power or energy is within: "From an invisible center in the mind we all become visible. What we have externally become, we have previously been mentally." She bases her simple theory on the concept of a center. "All things are created from the center and proceed outward. . . . Consciousness begins at a limited point and from this gradually awakens and becomes wider as it advances toward the center."[9] This conceptualization is reminiscent of Jung's model of the mandala, and is also compatible with Freud's ideas on the infinity within the unconscious.

Boheme writes that this "unchanging permanent energy" of the universe exists in our minds in thought. "Let your thought conform to the cosmic law. Let it rise from the symbol to the reality. . . . At the center, we are one with God, at the periphery, we are human."[10] As with

Boheme, Beers concludes that there is only one power in the universe from which all things proceed. He states that all of its manifestations can be reduced to three factors: matter, ether, and motion.[11]

These metaphysicians also discuss the so-called universal mind and the role of the ether and vibrational forces in understanding various forms of ESP. These were well-accepted esoteric hypotheses long before Jung's collective psyche appeared in the literature. In fact, Jung's original concept was quite clearly not the same as the universal mind. It was originally construed as a genetically passed-on individualized psychic structure. Only later—with his study of Swedenborg, synchronicity, and Buddhist philosophy—did Jung convert his theory to one that suggested that mind was all-pervasive. However, it was Freud's attitude, rather than Jung's, that was really responsible for the crumbling of the belief in the universal psyche. Freud's genius in analyzing and delineating the structure and mechanics of such an abstract realm, which he called the unconscious, was an unrivaled exploration. His theories so overwhelmed contemporary thought that the conceptualization of the divine origin of the human subconscious and its universal interconnectedness went the way of the passenger pigeon, even though his theories on the Oedipal complex are compatible with Jung's idea of archetypes existing in a collective psyche.

Freud's vehement negativity toward parapsychology, cited by Jung in his autobiography,[12] finally cooled sometime around 1920. However, the damage had already been done, and parapsychology has yet to recover from the influence of Freud's initial bias.

To this day the theory of the universal mind is held in low regard, although there is little doubt that Freud considered it. Ironically, Freud's writing not only paved the way for scientific scrutiny of the human psyche, but it also created adversity toward less famous explorers of the mind. The fact that Freud's *Interpretation of Dreams* never even considers telepathy as a viable cause of visions impressed upon the sleeping mind remains a subtle but powerful blow to the metaphysical aspects of psychology.

It is quite possible that this bias also influenced Jung. Even though he was open to parapsychology, he initially rejected or overlooked the theory of the universal mind when he developed his idea of a genetically inherited collective unconscious. Only later in life—when Jung began to see the collective psyche manifesting in synchronistic phenomena —did he become more open to speculating about such phenomena as telepathy. He also eventually came to adopt the idea of the universal mind in its occult form, but many Jungians to this day remain unaware of this.

OBSCURE OCCULTISTS

Other little-known but noteworthy thinkers (besides Beers, Coates, Hudson, and Boheme), circa 1900, were E. H. Anderson, a hypnotist from Toledo, Ohio; Richard Ingalese, a California lawyer; and Leonard Landis, a doctor from New York City. They all discuss similar hypotheses, each with a different emphasis.

Ingalese, a precursor to Ernest Holmes, of *Science of Mind* fame, was a colorful storyteller, his themes conveying powerful religious overtones. Interestingly, he remained critical of the Society for Psychical Research because "not one of the investigators has yet discovered the true cause of the phenomena or formulated the law under which they occur."[13] The law and cause behind the psyche's operation are the focus of his 1902 masterpiece, *History and Power of Mind*. He discusses ancient occultism, divine and dual mind, self-control, colors of thought vibration, dangerous psychic forces, and other laws of mind. His quaint esoteric anecdotes remain controversial but contain hidden truths. In describing the "dwellers of the threshold," the elementals, fairies, adepts, and denizens found when one enters the world of the astral plane, Ingalese writes:

There is still another kind of thought creations of man, which become embodied soon after they are born. These are the licentious

obscene thoughts of both sexes that become the creeping crawl-
ing bugs and vermin, which infest untidy homes, second and third
class hotels, and public houses. Then there are the biting stinging
thoughts, which embody themselves as flies, wasps, bees and mosquitoes, and the poisonous thoughts, which become spiders and reptiles. These miserable creatures born of man's lower mind, cannot
use the atoms of a higher rate of vibration for their bodies, but must
use those atoms, which they vibrate harmoniously with.[14]

Out of context, this passage can be easily misinterpreted, for what
Ingalese suggests is that "man himself created the destructive things of
the Earth." Although some passages are obviously dated, off-putting,
and involve magical thinking, nevertheless Ingalese's work remains full-
bodied, and has a side quite compatible with much of contemporary
thought. He discusses hypnotism, out-of-body experiences, and pre-
Freudian studies by Charcot, Bernheim, and others concerning a "secondary mind in man," that is, the subliminal self. He discusses the auric
double, its various colors, and its relationship to the inner character of
the person. He also states that character does not change with death,
and that many of the departed are still attached to the Earth plane,
still influencing the lives of the living through partial possession and
by means of weakness of will. Thus, prostitution, alcoholism, and other
forms of obsessions could have a psychic explanation. "When subjective
mind [personal unconscious], has incarnated into the objective mind
[universal] we have real psychic man, the dual man." However, man
remains ignorant of his divine linkage and he tends to live far below
his capacity. "The environment shows which consciousness controls. . . .
If you continue to create ignorantly you will suffer as though you knew
the law. . . . Both your minds can create good, but . . . objective mind
usually does not do so until it has been properly trained."[15]

As with Boheme, Anderson, and many other turn-of-the-century
psychic researchers, Ingalese deals with the role of the *will,* a psychic
function and structure virtually ignored by Freud and Jung and, of

course, Skinner but discussed at length in Ouspensky's and Gurdjieff's writings, which state that the highest states of consciousness are directly equated with acts of one's own willpower.[16]

Anderson's 1901 text, *Psychic Developments,* was discovered by Robert Adsit in a secondhand bookshop in New York City's Lower East Side. The size of the slim green text with gold lettering and lack of page numbers is misleading, for it contains a treasure house of information concerning "laws of mind."

"Everything," Anderson writes, "is created by inner action. All form has its mental base."[17] Anderson was, no doubt, a stage hypnotist and probably ran an esoteric society in Toledo. His table of contents includes chapters on the will, intention, confidence, hypnotism, healing, magnetism, thought transference, and telepathy. Less of a historian than Ingalese, Anderson is more direct in his concretization of the abstract structure of the psyche. He writes that all mind is amenable to suggestion:

> Man is all mind. He is one with the Universal Intelligence. He is expressing himself on the material plane. He builds his own dwelling place and thought is the bodybuilder. As the monad starts to function on the plane of individualism, it ceases to vibrate in harmony with the Universal Mind. It attempts to live for the self and antagonize all else. It is attempting to isolate itself from the course of all power . . . from the real self. . . . Man . . . has within himself all the powers and attributes of the infinite, limited only by his individuality. Get in tune with the infinite and allow the Universal Mind to express itself through him without any resistance from the conscious.[18]

Concerning telepathy, Anderson states simply: "[T]hought is vibration." If the subconscious is one with the universal intelligence, then "what I know all others know." Anderson sees the will as a resultant force, taking the line of least resistance. *Intention* is what must be cultivated in order to influence the blind will. *Silence* is one great factor in all psychic work: "Learn how to draw into your own silence. . . . Ideas, concise thoughts, and meditation are the essentials to concentration. Each must be free of flaws. . . . The goal is to be totally conscious of the self."[19]

Then intention masters all suggestions, controls will, and you are free. Besides concentration, meditation, silence, and intention, you need to cultivate confidence to undertake and accomplish whatever is desired. Confidence begets confidence, and this is "simply the recognition of our potentiality." This is really a delineation of the Kabbalistic credo "To Will, to Know, to Dare, to Keep Silent," but Anderson applies and explains this credo. A psychic healer must "[g]ive the suggestion of hope, energy, and life more abundantly. Make the person passive to the infinite power within. . . . The healer should feel within him the throb of joyous life." No fear, no doubt must enter the mind of the healer. As with Ingalese and Beers, Anderson sees disease as discord with the infinite. He emphasizes that all things exist potentially in the mind and that "[t]hought is a formative constructive force, or energy. By thoughts of health and strength we implant a suggestion, which, if nourished and developed, will assure us immunity from disease . . . Truth can only be sensed through the soul."[20]

Ingalese adds that the healer should "destroy the mental picture of the disease, which the patient holds in his matrix, [and] raise the rate of vibration of the afflicted area supplementing old elements with new ones." This is done through projection and concentration on the aura. "We must understand that the ether is universal; it cannot be excluded from any plane or place. This universal consciousness is, as its name implies, everywhere. It is important to remember this universality because in thought transference or in treating mutually a person at a distance you must realize that there is no separateness in mind."[21]

"The ether," Ingalese goes on, "is magnetic, attracting every particle to itself. Each person is a center [reflecting the whole] whose, task is to become a vortex of positive energy. We cannot help others until we become strong ourselves. We must have force before we can import it to others. . . . The ether is fluidic. This impartial sea of Divine Consciousness flows according to the impetus given to it both by Deity itself and man; and it moves in the direction in which it is sent. . . . There are currents of love . . . destructive currents . . . and you will be taught how to attach yourself [to positive currents] . . . and avoid destructive ones."[22] There is no friction in this ether, which obeys the law of karma (cause and effect). You get back exactly what you give.

A SKEPTIC

The late nineteenth century, however, was not without its skeptics. In a thin book entitled *The Philosophy of Spiritualism and the Pathology and Treatment of Mediomania,* 1874, by Frederich R. Marvin, M.D. (purchased for five cents), the author seeks to disprove the belief in an afterlife:

> I am informed there are four million men and women in America who believe in Spiritualism and whose minds are never lifted from its delusion. Men and women who, crazed with wonder at some trivial event, set aside the teachings of philosophy and common sense and face destiny with a lie. . . . Never do they lift their eyes in wonder but are wrapped in awe and transported with delight at the gyrations of a three-legged table or the incoherent raving of a crazy woman. . . .
>
> But if we are going to be deranged with the wonderful, let us have as healthy a derangement as possible. Let us go wild over the green and blue heavens; the stars that make the night beautiful and the Sun that makes the day golden; but in the name of taste and culture let us not select a tipping table nor illiterate phantom.[23]

Dr. Marvin's talk is persuasive and, in fact, his point is well made. We rarely appreciate the miracle of our very existence, let alone our psychic powers. Although the number of skeptical books on metaphysics is far outnumbered by the opposite, modern-day parapsychologists are all too aware that a single skeptic can virtually destroy the credibility of practically any psychic experience, whether spontaneously produced or created in the laboratory. Interestingly, it is statements such as Dr. Marvin's practical, but truly unscientific, remarks that lie at the bottom of the mysterious unacceptability of our higher nature.

PSYCHOANALYSIS AND BEYOND

Perhaps my most treasured find, again discovered by my friend Robert Adsit in Greenwich Village, is a work entitled *Psychoanalysis and Beyond,* by Leonard Landis, M.D., written in 1924. Landis was a clinical instructor at the University of New York Internal Medicine Department, editor in chief of the *Life & Health Magazine,* and national chairman of the American Association of Independent Physicians. As with James Coates, cited earlier, he was a member of the American Society for Psychical Research.

> Imbedded in the protoplasm of the cell is an upward, irrepressible climbing urge. Bergson has called it the *élan vital.* Like a purifying breeze the vital urge sweeps through the windings of evolution from worm to man. It is never-ceasing and underlying, pushing up through all forms to create higher ones. . . . The force that palpitated through the chain of life, I call the *center of perfection.* It rests at the heart of life, controls, directs, urges, and infuses creation with unity, purpose, ideals . . . It is the animating principle of Universal Intelligence often referred to as God. . . .
>
> The subconscious in man is directed by the center of perfection. . . . It is responsible for the strivings of poets, painters, scientists, and truth seekers of whatever nature. . . . It comes to our aid in times of stress and consoles us in the darkest hour. . . . In crowds, the center of perfection can almost be felt. . . . [It] is the consummate purpose running through all things, surmounting all things.[24]

*Fig. 2.3. Leonard
Lincoln Landis, M.D.*

In his chapter "Psychocosmology and Beyond Psychoanalysis," Landis writes that by research into the field of the subconscious and spiritualistic and physical domains, and by "cultivating the belief that the power and liberating vision of the soul are boundless as it is infinite, the accomplishment of the Divine is not a mission so much, which we have upon us, but a destiny, which cannot, even if we could, forego." All the world is conscious, but also it has a subconscious. "Psychocosmology agrees with Leibniz' monadology in that all perfection and accomplishment is in us in the nucleus, in their potentialities."[25] With the development of our supersenses "will come light and understanding verification of the perfection within and about us" and also the ability "not only to communicate mentally with other people at the far ends of the Earth, but with other universes at remote poles of space."[26]

He sees evil as a maladjustment of the perfection urge. This urge is in the will, not in consciousness, except as it acts as an instrument of this evolutionary impetus. The movement away from the origin is inevitably aimed to return to the ever-progressing source.

We are not ends, we are emanations. . . . Life is an irreversible wheel. . . . The urge to perfection is [for] the microcosm to become equal to and identical with the macrocosm; for the part to become whole. . . .

The physical can only know the cosmic through intermediaries. That is why consciousness, which operates only in the field of the physical, cannot know truth, except as it immerses itself or becomes identical with the subconscious. . . . What the physical eye sees, through its limited perception, the psychical eye *knows* [because it is one with its perceptions]. . . . The way to truth is only possible through the subconscious. Let us not forget this.[27]

"The way to truth is only possible through the subconscious." This is a thought worth repeating and reflecting upon. Landis's psychocosmology is an unparalleled work, for it adeptly integrates Freudian psychoanalytic theory with Jungian and esoteric philosophy while including a survey of his own soul and clairvoyant perceptions.

STORIES OF INFINITY

Some of these authors clearly write from experience, some from inspiration, but most derive their theories from other obscure and non-obscure psychic researchers and metaphysical thinkers who preach in parallel with every great civilization. Esoteric literature has permeated every segment of society from Bible groups to power of positive thinking insurance salesmen conferences, and their knowledge has aided everyone from Eileen Garrett to Exxon. Yet their work remains virtually unnoticed. That is why this tradition is truly hidden.

We will end with a quote from one last buried treasure, written in 1873 by the well-known French astronomer who later became president of the Society for Psychical Research, Camille Flammarion. The book is entitled *Stories of Infinity, Lumen—History of a Comet in Infinity*. It is a story about a man who meets a sagacious comet who has just returned to the Earth after traveling through the solar system.

There are some who, during life in the body, never lifted themselves toward heaven, never longed to master the laws of creation. Such,

Fig. 2.4. Camille Flammarion

still under the dominion of bodily appetites, dwell a long time in a state of trouble and unconsciousness. There are others, happily, who at the close of this life, soar on winged aspirations to the summits of sublime eternal; these behold with calm serenity the approach of the moment of separation; they know that progress is the law of existence, and that they will enter, beyond, into a higher life than this; they note, step by step, the numbness that mounts to their hearts, and when the last flutter, faint and imperceptible, ceases, they are already far above the body, which they had seen sinking into sleep; and shaking off magnetic bonds, they feel themselves borne of an unknown force toward that point in the creation to which their aspirations, their feelings, and their hopes have attracted them.[28]

Other obscure metaphysicians are hiding out at flea markets, antique shops, and equally obscure libraries, where other secrets are revealed. Our quest for parapsychological truths need not be limited to the archives of the British and American Societies for Psychical Research or to current laboratory experiments. This route has been traveled before. With compassion for yesterday's closet mystics, we modern-day truth seekers can learn much from the past, so that as we traverse new psychic realms, we will locate the sirens, steer clear of dangerous reefs, and light our way to tomorrow.

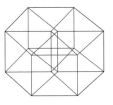

3

TOWARD A PHYSICS OF CONSCIOUSNESS

This chapter was first published as an article describing the "Toward a Physics of Consciousness" Symposium that was presented by Interface, of Watertown, Massachusetts, at the Harvard Science Center in Cambridge, May 6–8, 1977. Even though the article was originally published over a quarter century ago, the ideas presented at the conference are still on the cutting edge, as theoreticians look to quantum physics to help explain how consciousness operates. Present-day theories discuss: holographic and volitional properties of light particles called photons; Heisenberg's principle of uncertainty; so-called hidden variables able to bias this uncertainty; and the "tachyonic" realms, dimensions exceeding the velocity of light. Dramatic breakthroughs in the theories of physics are beginning to unravel the mysteries of the mind, and as such must be seriously dealt with.

The conference was an interesting conglomeration of career military personnel, inventors, and parapsychologists—all theorists in the new field of paraphysics. The speakers included Tom Bearden, NASA research scientist in Air Defense; E. H. Walker, Ph.D., Army thermonuclear warhead expert; Itzhak Bentov, biomedical and biofeedback engineer; Arthur Young, inventor of the Bell helicopter; Christopher Bird, coauthor of *The Secret Life of Plants;* and Andrija Puharich, M.D., psychic researcher and inventor of biomedical devices. Since that time, a number of the key speakers have become ill or passed away. Itzhak Bentov died shortly after the conference in an infamous Chicago plane crash that took the life of CIA operative E. Howard Hunt's wife. Christopher Bird, Arthur Young, and Andrija Puharich all passed away of natural causes in the 1990s, at the respective ages of sixty eight, ninety, and seventy five.

Before his death I came to know Andrija Puharich quite well, having interviewed him for *Gnostica* about a year after the conference. I then met with him several other times, mostly at New Age and Tesla conferences, and also spoke with him at length on the phone, as he helped me with my research on the inventor Nikola Tesla. Andrija was one of the most fantastic individuals I have ever known. On the one hand, he was simply a hard-nosed scientist studying the neurophysiology of telepathy, yet on the other hand, he saw himself as a liaison to an extraterrestrial group, the Eloheim, or the Nine, which he regarded as being involved in the genesis of human life on the planet.

For those who remember, there is little doubt that Andrija changed the world in the early 1970s, when he brought the paranormalist Uri Geller to the United States to be tested for his unusual abilities. A medical doctor, Puharich understood that if he worked with Geller alone, his research and findings would carry little weight. So he set up tests for the young psychic with some of the most advanced military think tanks in the country, and with other respected scientists around the world. Andrija was candid about his claims that he was in constant contact with extraterrestrials. In an offhand manner he informed me at his home in Ossining, New York, in 1978, that he had a watch that gave him information when "they" wanted to contact him. Ironically, his worldview was too radical even for Mr. Geller, and they parted company some time around 1975.

On the other hand, as a traditional scientist, Andrija had over fifty patents, most of them associated with novel developments in hearing aids, and his theories on the quantum physics of consciousness were well respected. He also had information on the inventor Nikola Tesla that I could not have obtained from any other source, because of his friendship with the inventor John Hays Hammond Jr., who had worked with Tesla shortly before World War I. Andrija and Hammond performed seminal telepathy studies with Eileen Garrett, the great psychic and founder of the Parapsychology Association. Working at Hammond Castle in Gloucester, Massachusetts, in the early 1950s, Puharich set up telepathy tests with Garrett locked in a Faraday cage, which screens out electromagnetic waves. Since she was successful in these tests, it established that ESP did not use conventional electromagnetic (EM) frequencies as a carrier wave.

It was Andrija who turned me on to the Tesla/Hammond connection, and it was also Andrija who introduced Tesla's patent application for a particle beam weapon to the world at an international Tesla conference held in Colorado Springs in 1984. This was a top-secret paper that had been hidden from public view for over half a century. Some time in the early 1980s, Puharich's home in Ossining was burned to the ground. Puharich claimed that the government was responsible, as he also claimed that he had worked in the past with agents of the government on secret work in parapsychology. I remember him telling me one day, before the fire, that the government was after him. I asked him how he could be protected, and he said he was like a patient being operated on in the dentist's chair with his hands around the dentist's testicles.

The other speaker I came to know well was Tom Bearden. An army colonel and war games analyst from Huntsville, Alabama, Tom was editing an esoteric newsletter called *Specula* and writing books such as *Excalibur Briefing.* Topics of these treatises ranged from the bizarre to very advanced theories on hyperspace weaponry systems, scalar wave technology, and the quantum physics of consciousness. His theories on esoteric aspects of Tesla technologies are quoted in my Tesla biography.

For the most part, Tom and I used to meet at the biennial Tesla conferences held between the years 1984 and 1996 in Colorado Springs.

In 1988, as we were both speakers, we were slated to share a room. However, there was only one bed, and we joked about it because Tom is the size of a large football player and I am not. A fellow overheard our plight and offered me the extra bed in his room.

Naturally, we got to talking. His name was Tim Eaton. Tim was a film editor who worked for Lucasfilm at Industrial Light & Magic, George Lucas's special-effects studio in San Rafael, California. Tim was looking for a Tesla screenplay and I had one, so we became partners, he on the West Coast and me on the East. About five years into the relationship I said, "How come you didn't wish me happy birthday?" "Why," he asked, "is this your birthday?" "Yes," I replied. "Well, it's my birthday too!" We have been creative partners ever since.

The most prestigious speaker was Arthur Young, inventor of the Bell helicopter. Parapsychology and the quantum physics of consciousness were controversial topics. Naturally, it was beneficial to the field to have someone of Young's stature. Co-editor, with mathematician Charles Musès, of the consequential text *Consciousness and Reality,* Young had also authored *The Geometry of Meaning,* a book that many see as an important synthesis of Pythagorean/Platonic ideas with those of present-day physics. With his wife, Ruth Forbes Young, who was the great-granddaughter of Ralph Waldo Emerson, Young had created the Institute for the Study of Consciousness in Berkeley, California.

The conference was organized by Ira Einhorn, a robust hippy from Philadelphia who had a long gray ponytail, wore flannel shirts, and sported a beautiful blonde girlfriend, Holly Maddox. Einhorn, as psychedelic philosopher, had become the liaison between the New Age movement and corporate America. Synergy was his calling card.

Shortly after this conference, Einhorn wrote to me with the hope that I would publish one of his poems in *MetaScience,* the journal I was editing. He was on his way to Europe, in part to visit the Tesla Museum, but shortly before he left, his girlfriend, Holly, disappeared. Einhorn made no secret of his contention that a faction of the government had kidnapped her. Most of his friends accepted Einhorn's accusation, particularly when Puharich lost his house. Einhorn was a charismatic character whose friends included a full range of people from counterculture figure Abbie Hoffman, to politicos in Philadelphia, to IBM and AT&T executives. He had succeeded in placing this conference at Harvard University, no small feat. Thus, Einhorn's accusations carried some weight, even when Holly's body was discovered, drained of all of its blood, in a trunk in a back room in Einhorn's apartment and he was charged with murder. People just thought, "That's how treacherous those secret agents are."

With the help of his lawyer, Arlen Specter, the latter-day Pennsylvania senator, Einhorn got his bail reduced to $4,000 cash. With his life in danger, he skipped bail and disappeared. It was only much later, with the 1990 publication of Stephen Levy's Einhorn biography, *The Unicorn's Secret,* that it became obvious that Einhorn had been lying, and that he had, indeed, murdered his girlfriend. But then it took more than another decade for the authorities to locate him. After he had been underground for more than twenty years, he was found in France, a white-haired, aging, thin shadow of his former self. His story became a front-page burner. However, because he had been tried in absentia, France blocked his extradition, and it took yet another year to get him to the States. Finally, in 2003, after a TV miniseries was made of the event, he was placed in an American prison, tried, and convicted of Holly's murder. However, this conference was long before all of this and, as you will see, it was a very exciting event.

IRA EINHORN AND CHRISTOPHER BIRD

The conference began Friday night with an introduction by Ira Einhorn on a "Planetary Overview." Einhorn, a longtime hippy and product of the antiwar, LSD culture of the sixties, discussed his own perplexity concerning the structure of the individual and society. Realizing not only the inadequacy of Western culture's inability to explain the mystical world he had entered, but also the need for a comprehensive and more accurate model of the universe, Einhorn had turned to Eastern philosophy and parapsychology. He stressed the need for a spiritual rebirth; he stated that we need "semi-permeable membranes" to achieve unity between polemic aspects of society, but that we must also turn to inner dimensions to reunite ourselves with the higher forces that created us.

Saturday morning the colorful Christopher Bird discussed the history of dowsing in a characteristic Irish brogue. He mentioned dowser Vernon Craig, who had stopped a drought in California by dowsing for water beneath the dried-up Lake Elsinore. To the amazement of authorities, billions of gallons were discovered and the lake was refilled. Although the county rejoiced and thanked the drillers, no credit was given to Craig.

Bird concluded with an anecdote regarding the work of researcher Dr. Bickel, who discovered that properties of resonance transmitted by ultrasound through the roots of trees and plants improve their growth in an extraordinary manner. As proof of these claims, Chris produced an already dried-out lemon that was still the size of a softball! He claimed that everyone he spoke to insisted it was a grapefruit.

ARTHUR YOUNG: INTENTIONALITY AND UNCERTAINTY

Arthur Young followed with a discussion of the theories found in his books: *Reflexive Universe, The Geometry of Meaning,* and the important compendium *Consciousness and Reality,* co-edited with mathematician Charles Musès. Young stressed the nature of intentionality, an area neglected by science, but that, in and of itself, is linked to the term "consciousness." He mentioned that the famous physicist Werner Heisenberg theorized that the photon, along with all other elementary particles (electrons, protons, and neutrons), has an indeterminate aspect. This is Heisenberg's principle of uncertainty, which holds that neither the velocity nor the position of an elementary particle can be determined with precise certainty. In other words, one of the very properties of the stuff that atoms are made of is indeterminate. This built-in uncertainty, diametrically opposed to Newton's precise clockwork model of the universe, lies at the basis of quantum theory, and states that the universe is built on a framework of probabilities, not exact certainties. This foundation reduces greatly any notion of predetermination and thus allows the idea of free will within the theory of quantum mechanics.

Since the building blocks of life are not totally predetermined, Young speculates that these blocks must have intentional aspects. Consciousness itself may permeate matter. He points out that Leibniz (and Aristotle) believed in the intentionality of the cosmos; Max Planck, one of the fathers of quantum theory, speculated that the photon might display purposeful behavior.

Young also described the enigma of the photon. The photon, being light itself, ironically is the "glue" that holds matter together. In his book *The Tao of Physics,* Fritjof Capra tells us that electromagnetism, the force that holds atoms together, is the sharing of photons by elementary particles. Physicists tell us that this tiny particle of light is massless and yet contains infinite mass. Therefore, Young says, it is at one with its goal. There is no distance for the photon.

Young described the famous experiment performed by Einstein, Podolsky, and Rosen that demonstrated that two light beams emitted in opposite directions from a common source "keep in touch with each other" in a way that cannot be explained by current theory. If the experimenter does something to one beam at a distance from the source, such as polarize it, the other light beam acts as if it knew what happened to its partner. The implication is that signals—but not necessarily energy—must traverse the intervening space, exceeding the speed of light. Physicists call this phenomenon of instantaneous information transmission "nonlocality."

According to Young, "Heisenberg's quantum of uncertainty makes free will possible. The energy for this is small, but mechanical devices, such as the lever, show that a small effort can have a large result. A single photon could blow up a city. The amount of energy is not the key. It is the hierarchy of distribution."

BEN BENTOV:
CONSCIOUSNESS AND THE COSMOS

Ben Bentov, author of the then just published book *Stalking the Wild Pendulum,* truly enjoyed himself as he lectured. Bentov stated that consciousness permeates everything and therefore even nonphysical entities evolve. The nervous system is the site where the nonphysical meets the physical. To explore the physiology of meditation, Bentov invented a seismographic device to record the reverberations in the aorta caused by the beating of the heart. During normal breathing, the reverberations in the aorta are out of phase with the heartbeat and the system is inharmonious. However, during meditation and when the breath is held, the echo off the bifurcation of the aorta (where the aorta forks at the pelvis to go into each leg) is in resonance with the heartbeat and the system becomes synchronized, thus utilizing a minimum amount of energy. This resonant beat is approximately seven cycles per second, which corresponds not only to the alpha rhythm of the brain but also to the low-level magnetic pulsations of the Earth.

Bentov also stated the Pythagorean view that the body acts like a musical instrument, each area responding to a different chord. As proof of this hypothesis, he interviewed meditators and found that a large percentage of them hear high-pitched frequencies. By testing them with hearing equipment, Bentov was able to label distinctly different frequencies, which he paired with various inner cavities of the brain. He concluded that these frequencies and their harmonics are an avenue of communication for the respective cavities, that sound and vibration are used by the brain for sonic forms of intra-cortical information exchanges.

Bentov hypothesized that the pituitary gland and the pineal gland communicate directly with each other by use of resonating frequencies across the third cavity of the brain—a standing wave connecting them. The pituitary gland is considered the "master gland," which controls all the other hormonal glands of the body. In kundalini yoga, it is considered the sixth or brow chakra. The pineal gland, the purported third eye, is the seventh or crown chakra. Compare this finding with Keith Floyd's that "the pineal body occupies the midpoint at the center of a neural energy field at which point occurs the burst of light that is experienced [holographically] as the screen of consciousness."[1]

TOM BEARDEN:
CONSCIOUSNESS AND TIME

Tom Bearden, a tall, heavyset, Hawaiian-shirted Texan, is a deep, precise thinker who knows his fundamentals. He discussed the cold war, his interpretation of UFOs, and his ideas about human perception. "Consciousness is time, specifically time delay," Bearden said. At particular points in time, the physical and mental worlds are identical. Humans, as products and reflections of the universe, are thus holographic miniature universes themselves. However, in some sense, what we are inside our mind and what is outside are the same! "Think a bit," he said profoundly, "a detector can detect only an internal change to itself, nothing else."

Ken Wilber says it this way, referring to the point Bearden is making and also the principle of uncertainty: "To measure anything requires some sort of tool or instrument, yet the electron weighs so little that any conceivable device, even one as 'light' as a photon, would cause the electron to change positions in the very act of trying to measure it! This was not a technical problem, but, so to speak, a problem sewn into the very fabric of the universe. These physicists had reached the annihilating edge, the assumption that one could dualistically tinker with the universe without affecting it, was found to be untenable. In some mysterious fashion, the subject and the object were intimately united and the myriad of theories that had assumed otherwise were now in shambles."[2]

Bearden also discussed the photon, which he said is a particle that makes a right-angle turn out of our three-dimensional world. It is three-dimensional, due to its orthorotation (spin); but from its vantage point, it "sees" our realm only as two-dimensional because it can do something (travel at the speed of light) that things in our physical world can't do.

Further, different moments in time are separated by photon interactions, the implication being that photons interface between dimensions. Like Arthur Young, Bearden was pointing to a relationship between light energy and consciousness. Just as photons escape this dimension, (because they travel at the speed of light), so does our mind because it too can transcend the present, but it can also reflect on the past or today, consider the future, and also observe or even act on the progression of time—this ongoing NOW—as physical time proceeds.

In reflecting on Bearden's ideas thirty years later, it also occurs to this observer that Bearden could have just as easily said that photons are four-dimensional objects that view our world as three-dimensional. One way or another, his point is that photons operate in a dimension one greater than our physical world and this realm is isomorphic (coalescent) with the realm of our consciousness.

Bearden then developed what he called the law of multiocularity, which was derived from the work of Aristotle. This "law" helps explain the classic enigma in quantum physics whereby elementary particles (electron, proton, neutron, and photon) act sometimes like waves and sometimes like particles. They are both monocular (particle) and multiocular (wave) because opposites can and do exist simultaneously. This logic—similar to physicist Niels Bohr's principle of complementarity, which states that an elementary particle can act like both wave and a particle—also allows a photon to have no mass and infinite mass at the same time.

Bearden's explanation of UFOs was based on a Jungian model of mind in that he combined the concept of the collective unconscious with current events: namely, the cold war with Russia. Since he has been in the military for many years and is involved with NASA and Air Defense, he is sensitive to the potential for thermonuclear war. Just as psychological pressures can cause archetypal images in a sleeping person to "pop out" into a dream, Bearden suggests that the UFO phenomenon may be a psychophysical materialization from the collective psyche. Bigfoot and the Loch Ness monster are also materializations from our collective psyche. Although in a Heisenbergian sense they are still not completely "determined" (that is, physical), the more we look for them, the more we are apt to create them.

ANDRIJA PUHARICH: TACHYONIC REALMS

Sunday morning began with the illustrious medical doctor Andrija Puharich. He discussed psychic healing, psychokinesis, teletransportation, and the Geller children. He showed films of Arigo, the Brazilian healer; table levitations; and the Stanford Research Institute's studies of Uri Geller. To explain psychokinesis, Puharich discussed various theories concerning resonant effects, extra-low electromagnetic frequencies (see John Taylor's *Superminds*), and tachyonic dimensions, realms that are faster than the velocity of light. Gerald Feinberg, physicist from Columbia and author of *What Is the World Made Of? Atoms, Leptons, Quarks and Other Tantalizing Particles,* had crystallized the concept of a tachyon, a theoretical particle that travels faster than the speed of light. Puharich said that the tachyonic realms, different harmonics above light speed, are responsible for the phenomenon known as psychokinesis (as well as all other mental phenomena). The brain steps these dimensions down to the physical plane. Later in the symposium, the writer-astrologer Louis Acker proposed a similar theory when he

Fig. 3.1. Andrija Puharich, 1984. Photograph by Marc J. Seifer.

briefly described his holographic model of mind, which includes the tachyonic realms, principles of higher harmonics of light speed, and laws of vibrations and their interactions. According to Acker, the mind, dwelling in the tachyonic dimension, steps itself down to the physical plane through the principle of resonance and harmonics.

Puharich also described incidents from his forthcoming book *Time No Longer* (which apparently was never published) concerning the so-called space kids he was then studying. To qualify for this distinction the child had to be able to psychokinetically bend metal. Puharich described one child who materialized a tree, another who claimed spiritualistic contact with Einstein, and a third who recited mathematic descriptions of the many parts of the electron. He said that six of these children had teleported to his place at Ossining, from as far away as Switzerland.[3] Although the children did not have control over this ability, Puharich stated that a meditative technique supposedly helped them. In it the subject visualizes a tunnel of light, then "rides the light beam," dematerializing to this dimension and rematerializing at the required location.

It was Puharich's view that these kids are in psychic communication with various higher-order civilizations. Puharich has mapped out about thirty different ones, which he described in his book *Uri*. He estimated the total number of Geller children (the next race of humans) to be about 1.5 million. All of them, he felt, had chosen or were chosen before birth to come to the Earth, each with a specific cosmic task. Geller obviously is the most outstanding of this group.

In disagreement with Bearden's view, Puharich believed that many UFOs have "nuts and bolts," and that they are flown by beings higher than humans.

E. HARRIS WALKER: PARAPHYSICAL MIND

The last speaker was E. Harris Walker. A low-key fellow who dresses in baggy pants and a flannel shirt, Harris is a physicist whose paraphysical model of mind is the most generally accepted one. He stated that physics does not deal with consciousness, yet consciousness is the central reality of our existence. According to him, consciousness is not a length or a mass. It exists yet is nonphysical. However, since consciousness relates to the physical world, it must involve some process of coupling internal states with external ones. Quantum physics, with its principle of uncertainty, allows for the possibility of this linkage. Walker repeated the adage that the very act of measuring elementary particles disturbs them; the observer cannot be separated from what he observes.

Walker stated that psychokinesis involves the observer's ability to alter the indeterminate aspect of the physical event in question: "If the observer can choose what he wants, he biases the uncertainty. This leads to a macroscopic result as in dice-throwing experiments." Quantum mechanics, Walker pointed out, does not tell you what is happening. It only tells you probabilities. "The observer actually changes the world when he looks at it, and at the same time he learns something." The most important aspects of his theory are explained in his chapter in *Psychic Explorations,* edited by Edgar Mitchell and John White. They concern the concept of

> . . . [t]he state vector, which is not a single state, per se, but an infinite number of states that represent solutions to the basic equation for the same complete set of boundary conditions. The state vector is the collection of all these states. In quantum theory the system is not in one of these states but in the totality simultaneously. The state vector provides the complete representation of the system. . . .
>
> In actuality we see one state out of the collection given by the state vector. Thus the process of measurement of the system is said to "collapse" the state vector onto one of its component states . . . [which] is determined only probabilistically. This probabilistic interpretation of quantum theory is called the Copenhagen interpretation.[4]

Mind, or consciousness, Walker speculated, contains "hidden variables" that, as Young pointed out, might be *intentional* attributes of the system. These hidden variables affect the state vector but they are not functions of space or time. They may involve information exchange through the property of nonlocality, mentioned earlier, or through holographic or resonant effects.

LIGHT INTERLUDE

Two superb films were shown at the conference during the lunch breaks. The first, *Prelude,* by Ron Haves, is an artistic, experimental film similar to the abstract sequences from the finale of Kubrick's *2001: A Space Odyssey.* The movie is a brilliant and primal experience with powerful musical accompaniment and visions of an endless tunnel, galaxies, snowflakes, and other amorphous and geometric shapes evolving one into the other.

The second film, *Healing Effects of Color,* was created by John Ott, who worked for Walt Disney filming sequences of flowers growing, recorded by time-lapse photography. Ott not only displayed spectacular footage of flowers bursting open, but on his own, he also recorded laboratory experiments and human studies performed with different frequencies of light. Ott found a relationship between the onset of various viruses and improper lighting. UV (ultraviolet) light, destructive to

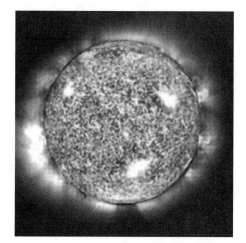

Fig. 3.2. The Sun with an infrared corona

some viruses, is needed to ripen an apple but not a tomato, as the latter utilizes light from the red end of the spectrum. He also noticed that a leaf's chloroplasts drop out of their natural tracks when there is an absence of UV light.

Strong blue light causes contractions of the rods and cones in the eye of a rabbit, and strong red light causes rupture and death of those cells. Ott found that X-rays caused violence in rats and sluggishness in humans. In one clever experiment, Ott placed plants in front of a TV screen, which was played six hours on weekdays and twelve hours on weekends. This was the national average viewing time for children. He found that their growth rate was stunted.

As a "photobiologist," Ott's work on the "therapeutic use of color" stemmed from Goethe's research, followed by such theoreticians as Rudolf Steiner and Max Luscher, both of whom realized that "the vibrational quality of certain colors is amplified by some forms, and that certain combinations of color and shape have either destructive or regenerative effects on living organisms."[5] Ott's experiments established that plants grow faster in red light, but, in the long run, grow sturdier in blue light.

His work is related to Kirlian photography, which is high-frequency electric photography that enhances the auras of such things as fingertips and leaves. Kirlian researchers had noticed that moods and disease affect the auras of people's fingertips. In one interesting experiment, the auras of the fingertips of two people who cared for each other blended when they kissed, but created a bar between them when they argued. Another experiment, cited later, in chapter 6, involved the phantom leaf effect, whereby the missing part of the leaf is still present as an aura when photographed this way. This research coalesces with that of Robert Becker, M.D., who has done extensive experiments on the link between the process of regeneration (as in the tails of lizards) and electrical fields. Becker has found that certain frequencies of electricity can be used to cause bones to knit in people who have trouble in this regard. And a third area involved the realization that Petri dishes growing various cells could be photographed with Kirlian photography after being given a variety of medications. The aura of the cells in the dish would be either enhanced or diminished, depending on whether or not the drug was beneficial. This process could speed up test trials of drugs on critical patients to the point of almost being an instantaneous test whereby normal procedures could take, several days when such a patient may not have those days to spare.

Royal Rife and the Cure for Cancer

Now, we jump ahead twenty-three years, to 2001 and a Tesla conference I spoke at in Tempe, Arizona, organized by Steve Elswick, editor of *Exotic Research* and *Extraordinary Science* and past director of the International Tesla Society, which was situated in Colorado Springs from 1983 to 1997. One lecture I attended discussed the importance of a blanket that gave off red light as a healing effect for injured horses. Another talk and demonstration, by Lynn Kenny on Royal Rife, took my breath away, for Rife, it seems, had found, more than fifty years ago, that light therapy could be used as the cure for cancer.

Up until 2001, I had never heard of Royal Rife, even though Christopher Bird had written an article on Rife in 1976 in *New Age Journal* (which I subscribed to at the time). But in 1996, while speaking at the Seventh Biennial Tesla Conference, held at Colorado Springs, I had stumbled upon Neil Gerardo, a mysterious fellow who gave an evening lecture on the cure for all diseases. Gerardo had a simple idea: If one could find the resonant frequency of a particular disease, such as AIDS, one could use an electronic gun of sorts to zap a person who had this disease and thereby destroy all aberrant cells in nanoseconds. He called his device MRX. The military, according to Gerardo, had such a weapon, which involved magnetic resonance and spectral-coupling X-ray technology, but was not using it for beneficial purposes. Because of the pharmaceutical lobby and pressure from the military-industrial complex, Gerardo claimed that he had to take his device to the South American country of Colombia to begin trials.

At the 2001 Tesla Conference I met Lynn Kenny. With him at his

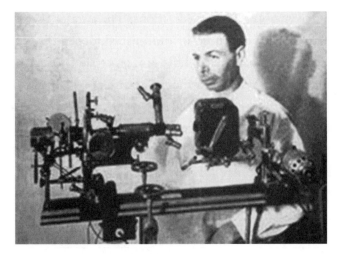

Fig. 3.3. Royal Rife and his universal microscope. With six thousand working parts, including a variety of lenses and prisms, and the ability to focus in three planes, the microscope could magnify 60,000 times. This allowed the viewer to study, for the first time, living viruses.

booth was a woman who claimed that Kenny had cured her of breast cancer with a Rife machine. This device was essentially a fluorescent bulb hooked to a computer that produced light rays of various specific frequencies targeted as detrimental to corresponding lethal diseases. They even displayed photos of the diminishing tumor to prove their success. The idea was simple and, as far as I can see, fundamental.

Rife (1888–1971), a scientist who worked for Zeiss Optics, had hypothesized as far back as the 1920s that cancer was caused by viruses, and that bacterial infections started from viruses as well. The problem in studying viruses, which exists even today, with powerful electron microscopes is that in order to view one, it must be magnified at least 40,000 times, and to do that, one must stain and therefore kill it. Because of this, viruses could not be studied in their natural environment. Rife wanted to view living organisms, so he set out to design an optical microscope that worked by using light, refraction, and resonant frequencies.

In the early 1930s, Rife completed his "universal microscope," a device that comprised more than six thousand parts and did not kill the organisms being studied. Able to magnify up to 60,000 times, Rife became the first person to study living viruses. The microscope worked by simply hitting the virus with different specific frequencies of light. Eventually, when the explicit frequency of a virus was obtained, the cells began to luminesce, and Rife was then able to focus his lenses to view these very tiny, dangerous, living organisms.

The key here is that Rife then knew the resonant frequency of the virus or cancer cell in question. This specific rate of vibration was charted and cataloged. Using a simple principle in physics, Rife understood that if he overloaded the cell with that same frequency, it would eventually disintegrate and wither away. Keep in mind that all molecules are held together by light energy, as elementary particles share photons. This fundamental principle was memorialized in a famous TV commercial using sound waves instead of light for Memorex tape recordings, whereby Ella Fitzgerald's voice was used to shatter a wineglass when she hit the correct note. This mechanism is the cure for cancer.

A similar idea was patented a few years ago by Michael Nyberg, a high school student from Old Lyme, Connecticut, who used the concept as a nonpoisonous way to kill mosquito larvae. Nyberg writes, "I discovered a sound frequency, which matched the natural frequency of a mosquito [larva's] air bladder. When this sound is put into water it causes the air bladder to rupture, killing the larvae [*sic*]. Acoustic power levels are very low and the technique appears to be highly specific."[6]

Destroying cancer cells and other diseases, however, is more complicated. For instance, Rife understood that tuberculosis involved not only a particular bacterium but also a corresponding virus that went undetected. If a single frequency was used to just kill the bacteria, the patient would eventually die because the virus was still alive. However, when the two electronic frequencies were used simultaneously to kill both the bacteria and the virus, "the patient gets well."[7]

Throughout the 1930s and forties, Rife performed numerous

successful studies with mice that had tumors, and he also cured many people who were literally dying of cancer. Other doctors had success as well, and these results were published in the 1930s in *Science, Popular Science, the New York Times,* and the *Pacific Coast Journal of Homeopathy.* Rife's 1953 paper, "The History of the Development of a Successful Treatment for Cancer and Other Viruses, Bacteria and Fungi," was also cataloged in the medical library of the National Institute of Health, where it was ignored by cancer researchers for decades.[8] Unfortunately, since the cure was so simple, the medical and pharmaceutical lobby began to paint Rife with the same brush that was used against snake-oil salesmen, and his work was suppressed. Rife's story is outlined in detail in the must-own book *The Cancer Conspiracy: Betrayal, Collusion, and the Suppression of Alternative Cancer Treatments* by Barry Lynes.

Clearly, Rife's microscope should be reconstructed, manufactured, and made available to universities and other think tanks. An easy way to start testing the efficacy of Rife machines would be for university labs to simply purchase them (they are available online) and do animal experiments using the frequencies provided by the manufacturer. A better way, of course, would be to replicate Rife's work with animal and human experiments so that the resonant frequencies of various diseases could be cataloged and hopefully cured. The cost is minimal, the benefits to humankind—incalculable.

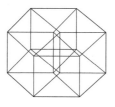

4

THE BIRTH OF THE
NEW PHYSICS

*It followed from the special theory of relativity that mass
and energy are both but different manifestations of the same
thing—a somewhat unfamiliar conception for the average mind.
Furthermore, the equation E is equal to mc squared, in which
energy is put equal to mass, multiplied by the square of the
velocity of light, showed that very small amounts of mass may
be converted into a very large amount of energy and vice versa.
The mass and energy were in fact equivalent, according to the
formula mentioned before. This was demonstrated by Cockcroft
and Walton in 1932, experimentally.*

ALBERT EINSTEIN

As the nineteenth century transitioned into the twentieth, explorations
in the field of physics radically changed scientific understanding of the
nature and structure of both space and time, as well as revealed the
importance of the observer. A review of the development of the physicists' view of the relationship of time to the macroscopic world (that is,

the structure of Einsteinian spacetime) and to the quantum physical realm (that is, the structure of atoms and elementary particles) provides a helpful background for our attempt to gain a more comprehensive understanding of the structure of time and the importance of mind in the spacetime continuum.

ETHER AND SPACETIME

Einstein's original theory of relativity stemmed from the nineteenth-century discussion of the so-called ether, the medium through which light traveled, and the inability of scientists to prove its existence. Isaac Newton (1642–1727) made use of the ether as a medium to explain gravity's action at a distance, but "whether this agent be material or immaterial, I have left to the consideration of my readers."[1]

When James Clerk Maxwell (1831–1879) formulated his electromagnetic theory, circa 1860, he saw magnetism as vortices in the ether. He hypothesized magnetic lines of force occurring at right angles to the direction of a flowing electrical current. But he "found great difficulty in conceiving of existence of vortices in a medium, side by side [that is, the magnetic lines of force] revolving in the same direction about parallel axes." He noted, "The only conception, which has at all aided me in conceiving of this kind of motion, is that of vortices being separated by a layer of particles, revolving each on its own axis in the opposite direction of the vortices."[2] Maxwell likened these vortices (which bear a striking resemblance to Penrose's twist-ors, discussed in chapter 11) to a well-known physical mechanism called an "idle wheel," which is placed between two other wheels intended "to revolve in the same direction."[3] Maxwell's conclusion is quite clear: "There can be no doubt that the interplanetary and interstellar spaces are not empty, but are occupied by a material substance or body, which is . . . the most uniform body of which we have any knowledge."[4]

The definitive experiment for the discovery of the ether was performed in the 1880s by Michelson and Morley.

The apparatus was mounted on a stone slab fixed upon a wooden float revolving in a tank filled with mercury, and made one full revolution in six minutes. A ray of light from a special lamp fell on mirrors attached to the revolving float and partly passed through them and partly was reflected, one half going [to a mirror situated twenty-two miles] in the direction of the movement of the Earth and the other [to another mirror twenty-two miles] at right angles to it. This means that in accordance with the plan of the experiment one half of the ray moved with the normal speed of light and the other with the speed of light plus the rotation of the Earth. [After bouncing back, the return rays would meet] at the union of the divided ray, [where it was hoped] there would have appeared, according to the plan of the experiment, certain light phenomena resulting from a difference in speed and showing the relative movement between the Earth and the aether, that is, proving the existence of the aether.[5]

An analogous situation would be to sit on top of a moving car and throw out two balls at equal speeds at right angles to each other. Obviously the motion of the car moving on the road through the air will affect the trajectory and ultimate speed of each of the balls. In the same sense, it was hypothesized that if the ether existed, it would affect the two different light beams. Yet, they both arrived back at the same time. Since no substantive differences were detected, the conclusion drawn was that either the ether did not exist or that by its very nature it could not be detected. "This seemed conclusive, but it had the embarrassing consequence of depriving electromagnetism of a most successful theory and leaving nothing in its place."[6]

Maxwell's ideas about electromagnetism had stemmed from the

work of Michael Faraday (1791–1867) and his predecessor Hans Christian Oersted (1777–1851), who looked at the structure of the ether in slightly different ways. Oersted had realized that when electricity is sent down a wire, a magnetic field is created around the wire. This combination of electric and magnetic forces created "a circulation action" that had polarity. These two forces, which interact in conflicting ways or in harmony, could thereby spread out to cover all of space.[7]

Oersted hypothesized that the nature of this interaction of two fundamental forces could account for five separate phenomena: electricity, magnetism, light, chemical action, and heat. He considered that all space is filled with a patchwork of forces "which manifest themselves according to the various conditions that exist locally in any state."[8] Michael Faraday expanded on this idea. The ether, for Faraday, was really "a three-dimensional web of lines of force crisscrossing to infinity."[9] Light would thereby be carried as a vibration along these energy vectors. Influenced by Oersted's work, Maxwell recognized the role of spin. Note that in this instance, the ether is not a "substance" so much as a complex medium of transfer.

Early-twentieth-century physicists found it difficult to reconcile these views with the constancy of the speed of light and its link to the Earth traveling through this hypothetical medium. If the Michelson-Morley experiment was correct, then it seemed that there was no ether to be detected. However, according to Price and Gibson (1999), the minor discrepancies noted in the famous experiment were indeed proof that ether did exist; it was just that "the relative ether velocity that they found was much lower than they anticipated."[10] The generally accepted conclusion, however, is that these differences—which range from .01 for Michelson-Morley in 1887, to .002 for Kennedy in 1926, to .0004 for Illingsworth in 1927[11]—are negligible. But are they? In any case, the implications of either a nondetectable or a nonexistent ether created new headaches for those attempting to resolve problems with mass and provide an explanation of how light traveled from the Sun to the Earth.

The Lorentz-Fitzgerald contraction theory was an early effort to explain the unexpected findings of the Michelson-Morley experiment. It proposed that matter contracted in the direction of its motion by a factor just sufficient to explain the negative results of the Michelson-Morley experiment. . . . As an illustration of this theory the length (L_1) of a body in motion would be given by the formula:

$$L_1 = \frac{L_0}{\sqrt{1-c^2/v^2}}$$

where L_1 = new length

L_0 = old length

v = velocity of moving body (in this case the measuring device moving with the Earth)

c = velocity of light[12]

This formula was used by Einstein in his 1905 special theory of relativity as an "instrument to translate data from one reference system to another in a state of uniform motion with respect to the first."[13] But the idea of length contraction being caused by the ether was disregarded. Einstein concluded that the ether, if it did exist, could not be detected. His theory also stated that there was no absolute frame of reference and that the speed of light is always constant with respect to faster- or slower-moving bodies. Later, Einstein's general theory of relativity (1916) was formulated to take care of accelerated motion. *Where the special theory linked space to time, and electricity to magnetism, the general theory combined these two couplets together.*[14] Because Einstein's theories were so powerful, this inadvertently led to a disbelief in the reality of the ether, although this was not his original intent. And because of this, scientists have ignored the ether ever since.

We can see, however, that Einstein's theories are limited to the extent that no definitive statement concerning the properties of the nineteenth-century ether are suggested. Therefore, the questions raised by Maxwell's view of the ether as a medium for light to travel through and Newton's suggestion that it also was the medium for gravitation are still basically unanswered, whether light is a wave, a particle, or a combination of the two.

It is interesting to note that the electrical inventor Nikola Tesla, whose key work predated Einstein's theory, assumed the presence of ether; because of this, he was able, in the early 1890s, to transmit electrical energy through the air and to light up fluorescent and neon vacuum tubes by means of high-frequency alternating currents and wireless power transmission. Tesla showed that when a vacuum tube resonated at a particular frequency, it emanated light, whether or not it contained a rarified gas or a filament. Tesla simply removed the filaments from Edison's vacuum bulbs and illuminated them as well. The vacuum, Tesla said to the dismay of his rival Edison, was more important than the filament.

> It is certainly more in accordance with many phenomena observed with high-frequency currents, to hold that all space is pervaded with free atoms, rather than to assume that it is devoid of these. . . . Is then energy transmitted by independent carriers or by the vibration of a continuous medium? This important question is by no means as yet positively answered. But most of the effects, which are here considered, especially light effects, incandescence or phosphorescence, involve the presence of free atoms and would be impossible without these.[15]

It is important to note that Tesla made these remarks a few years before one of his cohorts, J. J. Thompson, discovered the electron. Tesla's use of the term "free atoms" refers to the structure of the Maxwellian ether, a sea of spinning vortices. Tesla's idea that energy may be transmitted by the vibration of independent carriers resurrected Newton's treatise on optics in which the dual nature of light was first introduced. Tesla's work—presented to physicists like Nobel Prize winner Robert Millikan and Elmer Sperry, inventor of the gyroscope, at a lecture at Columbia University in 1889, and in the early 1890s to members of the Royal Academy of Science like Lord Kelvin, Sir William Crookes, J. J. Thomson, and Nobel Prize winners Lord Raleigh and Sir Ernest Rutherford—predated in concrete form the similar wave/particle theory put forth by the quantum physicists about a decade later.

Einstein, not totally secure in his theory of relativity, told Sir Herbert Samuel that "[i]f Michelson-Morley is wrong, then relativity is wrong."[16] Obviously Einstein realized that if the ether could be detected, then relativity was in error. Six years before his paper was published, in 1899, while he was still a student of mathematician Hermann Minkowski, Einstein attempted "to construct an apparatus, which would accurately measure, the Earth's movement against the ether," according to Rudolf Kayser, Einstein's stepson-in-law, who went on to say that Einstein never built the apparatus "because the skepticism of his teachers was too great."[17] We can therefore deduce that peer pressure played a role in Einstein's theory of physical reality, which in essence denied an ether he knew was most likely there!

Common sense, logic, and the lesson of history would suggest that in time Einstein's theory must be proved wrong in this respect, as eventually the ether will be detected. Yet, part of the problem is really one of semantics, for what Einstein really did was replace the nineteenth-century, three-dimensional Euclidean ether with a twentieth-century, four-dimensional non-Euclidean spacetime continuum, which had its own medium.[18] This conceptualization had shortcomings too, because it ignores the self-evident fact that space cannot be empty. It certainly contains the intersecting light rays from all the stars and the crisscrossing forces that Faraday and Oersted describe.

It should be noted that within the last twenty-five years, physicists claim to have detected an all-pervasive background radiation stemming from the so-called big bang, that is, the initial explosion that gave birth to the universe. However, this does not necessarily suggest that this is the same ether described by Oersted, Faraday, and Maxwell. There may, in fact, be a hierarchy of spaces, or ethers. The lake is not calm.

IMAGINARY NUMBERS, RELATIVITY, AND QUANTUM THEORY

Ronald W. Clark's biography makes it clear that Einstein owed much of his success to his college math teacher Hermann Minkowski (1864–1909). "Whether Einstein would ever have done it without the genius of Minkowski we cannot tell," said *Relativity* author E. Cunningham. "Minkowski gave a mathematical formulism to what had been a purely physical conception of special relativity."[19]

> The moment we ask ourselves if there are any other *kinds* of numbers but the ordinary ones, we open the door to an immense journey and the means of traveling it. Numbers are then to be seen as powers of transformation.[20]

Minkowski had won the Paris Prize in Mathematics at age eighteen. He was a hero for Einstein, as he was able to assimilate Riemann's

Fig. 4.1. Hermann Minkowski

non-Euclidean geometry used for curved and elliptical spheres with relativity, and provide equations for relativity, which allowed, in Einstein's words, "the time coordinate to play exactly the same role as the three space coordinates."[21]

What Minkowski did was introduce the mathematical unit $\sqrt{-1}\ ct$ (the square root of negative one times the speed of light times time) as the time coordinate; this in turn could relate symmetrically to the three space coordinates (x, y, and z, or height, width, and depth). An important point to keep in mind is that $\sqrt{-1}$ is an *imaginary* unit, that is, it has no physical counterpart; there is no square root of a negative number because a negative number times a negative number, by definition, always yields a positive number. Therefore $\sqrt{-1}$ = the imaginary unit. Yet the use of this imaginary number explained quite efficiently the mathematical formulism for relativity, and was intimately involved with the reasons why Einstein received the Nobel Prize.[22] By introducing a time-dimension equivalent to the three space coordinates, a "happening in 3-d space physics becomes, as it were, an existence in the 4-dimensional world."[23]

In 1903 Minkowski addressed a scientific convention in Cologne:

Gentlemen! The ideas on space and time, which I wish to develop before you grew from the soil of experimental physics. Therein lies their strength. Their tendency is radical. From now on, space by itself and time by itself must sink into the shadows, while only a union of the two preserves independence.[24]

Unfortunately, a year later Minkowski died of a burst appendix; he was only forty-four. At the same time, the European scientists, heavily influenced by Minkowski's mathematics, invited Einstein (along with Marie Curie) to Geneva to be presented with an honorary doctorate.[25]

The square root of negative one, which is an imaginary unit, is also called a *hypernumber* (an imaginary number—one that could not exist in reality). It has been used by mathematicians for hundreds of years and was adopted by Gauss in the early 1800s for use in non-Euclidean or non-flat geometry. The use of $\sqrt{-1}$ occurs in a realm one dimension higher than Euclidean geometry. W. R. Hamilton, a contemporary of Maxwell, found in 1843 "higher orders of $\sqrt{-1}$, which no longer followed commutative arithmetic and reacted differently to multiplication from the right or from the left. . . . Hamilton's friend and fellow mathematician, John Graves discovered still higher orders of $\sqrt{-1}$."[26] Having attended Hamilton's lectures, the brilliant mathematician Arthur Cayley (1821–1895) discovered a year later "a full blown non-associative algebra": for example, $(xy)z \neq x(yz)$. Studying the works of Lobachevsky and Riemann on such topics as the space on a curved surface, Cayley developed his own ideas on matrices, non-Euclidean geometry, and space with n-dimensions.[27] This nonassociative algebra creates an asymmetric framework compatible with the changing vector of time.

Science editor and mathematician Charles Musès states in his subchapter, "Hyperstages of Meaning," that numbers, as a language and description of reality, "are then seen to be powers of transformation."[28] The introduction of the hypernumbers, he adds, "more than doubled the entire mathematical power of all previous centuries. . . . They became intimately related to the new physical discoveries in electronics, atomic theory, and twentieth-century chemistry."[29]

Shortly after the discovery of the electron by J. J. Thompson, Niels Bohr published his theory on the structure of the atom. Bohr, much like Tesla before him, likened the electrons' paths, with circular orbits around the nucleus, to the motion of the planets around the Sun. However, this initial theory did not completely correlate with the observed frequencies in the line spectra of the hydrogen atom.

The prolific writer and Russian physicist George Gamow (1904–1975) describes the events he was privy to in his classic book, *Thirty Years That Shook Physics* (1966). Gamow, whose own work led to a better understanding of nuclear fission and fusion theory, informs the reader that it was the German physicist Arnold Sommerfeld who "extended Bohr's ideas to the case of elliptical orbits";[30] this addition to the theory of the structure of atoms better explained the motion of the electron. Wolfgang Pauli continued the study by formulating the Pauli principle, which distinguishes three quantum numbers to correspond to:

1. the diameter of the electron's orbit
2. the eccentricities of the orbit (azimuthal shape of orbit)
3. the space orientation of the orbit (orientation of angular momentum)

However, the data from the line spectra were still not completely satisfied.

Studies of the Zeeman Effect (the splitting of spectral lines by strong magnetic fields) revealed that there are more components than the three integers could account for, and to explain their existence, a fourth quantum number was introduced. . . . In 1925 two Dutch physicists, Samuel Goudsmit and George Uhlenbeck, made a bold proposal. This surplus line splitting, they suggested, is not due to any additional quantum number describing the electron's orbit but to the electron itself.[31]

Goudsmit and Uhlenbeck suggested that the electron was spinning upon its own axis. This orthorational component became the fourth quantum number. Now an electron's orbit could be described by clear parameters.

However, given that the electron was a particular diameter (3×10^{-13}), "it turned out . . . that in order to produce the necessary magnetic field, the electron would have to rotate so fast that the points on its equator would be spinning at much higher velocities than light!"[32] From the mathematical point of view, Gamow tells us, the anomaly was quite all right. However, this orthorotational velocity violated

Einstein's theory of relativity, although it violated no principle in quantum mechanics.

Paul Dirac (1902–1984) pondered this problem, seeking to create a quantum mechanical mathematics that would be compatible with relativity. Faced with a similar problem as the one solved by Minkowski, but now on a subatomic scale, Dirac was trying to create an equivalence between the time and space coordinates for the spinning electron.

Minkowski's use of $\sqrt{-1}$ allowed an equivalence between space and time on the macroscopic scale, thereby solving the problem of creating symmetry in the three space coordinates as compared to the time coordinate. Essentially the same procedure was used by Dirac to solve the problem of violation of relativity for subatomic particles. By introducing $\sqrt{-1}$, quantum physicists could create a formula that did not result in the anomaly of the orthorotating electron spinning faster than light speed.

Dirac used the same mathematical derivatives of each dimension, that is, either x, y, z, and t or x^2, y^2, z^2, and t^2, to account for the Pauli Principle. By using the four quantum numbers to describe the orbiting electron, Dirac not only succeeded in uniting quantum mechanics and relativity, but was also able to formulate equations that explained the remaining missing links in the split spectral lines of the hydrogen atom.[33] Dirac was then able to see the electron as acting like a tiny magnet. His equations and speculations also discussed the possibility of associating the "holes" in the atomic shells (which electrons had jumped out of) with anti-electrons having a positive charge.

Dirac's initial hypothesis also speculated about ether, which he suggested was composed of a sea of negative energy waiting to pop into vacant electron orbital shells. He saw this realm as being structured in a similar way as the physical one. If this were not true, then the stability of the universe would be threatened, for electrons in positive energy states would simply jump into negative energy states, thus annihilating the physical universe! However, if these negative energy states were already occupied, then this could not occur.[34] Dirac thereby resurrected the ether, but completely redefined it.

> Dirac suggested the existence of negative energy states of the electron. To interpret these states, Dirac postulated that empty space, the vacuum, was not really empty but filled with electrons in negative energy states but still in accordance with the Pauli exclusion principle . . . space is thus a three-dimensional ocean in four-dimensional space-time. What is below the surface is not normally observable. However, high-energy photons can occasionally knock holes into this surface. As a result, the "undersea" electron will leap out of the sea and will become a normal electron with positive energy and mass. But there is now a "hole" or bubble, left in the sea where the electron has been. This hole is a negation of negative mass: it will have positive mass and charge. . . .
>
> But such a particle, Dirac predicted, would be short-lived. Very soon, a normal electron would be attracted by the "hole" and fall into it, and the two particles would "annihilate" each other, dematerializing in a flash of high-energy photons. Thus, electron-positron pair creation and annihilation had been predicted.[35]

Once Dirac created an elegant solution for the spinning electron, which involved the imaginary unit $\sqrt{-1}$, the problem of violating relativity was neatly sidestepped. But George Gamow, one of the giants of quantum physics, in reviewing Dirac's accomplishment alludes to a huge secret that has been overlooked, or rather ignored. This text suggests, as many people intuit, that tachyonic realms must exist, and Gamow inadvertently gives us one proof: he informs us that if calculated one way, electrons orthorotate 1.37 times the speed of light (Sommerfeld's number). In my opinion, physicists took the easy way out, not because they introduced the imaginary number to deal with the problem, but because they ignored the fact that, according to their own mathematics, electrons spin at speeds in excess of the speed of light. What does this mean? I suggest that we simply take this information at face value. As we will see in chapters 10 and 11, it suggests that elementary particles, electrons in particular, interface across dimensions, between our so-called physical space and some hyperspatial realm.

TODAY'S ATOM

The original design for the structure of the atom consisted of four elementary particles: the electron, proton, neutron, and photon. Electrons, which are negatively charged (–), orbit the nucleus. They are fundamental, and appear not to consist of anything smaller. Inside the nucleus are protons (+) and neutrons (+ –). The latter ones are neutral, because they in turn are made up of a proton-electron pair. Photons, which have no mass and no charge, are tiny wave packets of light, which are used to bind one atom to another. In that sense, photons can be seen as the glue that holds molecules together. They are bundles of energy determined by the wavelength of light. When photons collide with electrons, they can act like solid bodies and knock the electrons out of orbit. Visible and UV wavelengths of light are too large to collide with electrons. Since neutrons are made from protons and electrons, one could say that according to this model, all matter consists of just two major building blocks, one with a positive charge and the other with a negative charge, along with the photon, which is the glue and carries no charge.

This overarching model was further refined after numerous studies were conducted with particle accelerators used to smash these four entities into each other. The present model of the atom is constructed of fermions, which are particles, and bosons, which are binding forces.

1. There are two kinds of fermions: quarks (with six subtypes) that form the protons; and leptons, which are electrons (and other particles, such as muons, tau, and two kinds of neutrinos). Note that protons can be broken down further (into quarks), whereas electrons are still seen as fundamental.

Fig. 4.2. The atom

2. Bosons are the binding forces; three are known and one is hypothesized. These four bosons account for the four known forces in the universe: a) electromagnetism, b) strong nuclear force, c) weak nuclear force, and d) gravity. The photon, as stated before, holds atoms together (a); gluons hold the nucleus together (b); the Z and W bosons hold neutrons together and (c); the graviton, also called the Higgs boson, which is a theoretical particle associated with gravity, gives fermions their inertia or mass.[36]

An examination of the sizes of the particles makes it clear that an atom, and therefore all of matter, is made up mostly of space: the proton is 1,836 times the size of the electron, yet 100,000 times smaller than the size of an atom. The first atom from the periodic table of elements, the hydrogen atom, consists of an electron orbiting a proton/nucleus. If this atom were the size of a room, the nucleus would be so small it would be invisible. If the atom were the size of St. Peter's Cathedral in Rome, the nucleus would be the size of a grain of sand; if the atom were the size of the planet, the nucleus would be the size of a three-quarter-acre lot.[37]

IMAGINARY NUMBERS

As we have seen, the use of imaginary numbers by Minkowski, Einstein, and Dirac led to better explanations concerning the structure of matter by providing for:

1. a unification of space and time
2. a unification of quantum physics and relativity
3. further discoveries concerning the orthorotational properties of the electron
4. the discovery of the anti-electron

Charles Musès writes that although numbers have become indispensable to life (such as phone numbers and zip codes), "numbers remain mysterious."[38] Hypernumbers beyond $\sqrt{-1}$ are crucial for the introduction of the mental realm into a mathematical model of reality, he says.

> The persistent and fruitful human intuition that numbers conceal powerful mysteries of comprehension and insight will be more than justified as men work with new and higher kinds of numbers, the hypernumbers. . . . New powers will thereby be released not only adding to man's ability to do things but also increasing his ability to heighten, deepen and illuminate awareness beyond what he previously imagined was possible.[39]

One of Musès's associates, the Nobel Prize–winning physicist Eugene Wigner, states that the imaginary number i or "$\sqrt{-1}$ enters quantum physics as a physical fundamental and not merely an elegant tech-

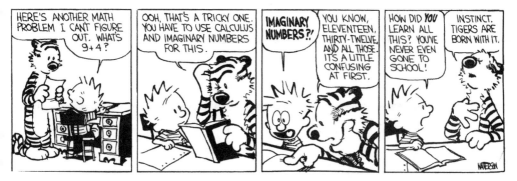

Fig. 4.3. Hobbes solves the mystery of imaginary numbers (from The Authoritative Calvin and Hobbes, *by Bill Watterson)*

nique."[40] Hypernumbers, which exist only in the mind as imaginary units, are necessary for explaining physical phenomena. Their role cannot be overemphasized, and it is more than likely that their use will eventually be expanded to explain functions of consciousness as well. The concept of hypernumbers is also intimately linked to hyperspace, which could be considered as a fourth dimension for space (for example, inner space or mind). If that were the case, time would be moved up to the fifth dimension.

KALUZA, KLEIN, AND WEYL

By the late 1800s . . . it was widely believed that in addition to ordinary material objects, there must throughout space be a fluid-like ether with certain mechanical and electromagnetic properties. And in the 1860s it was even suggested that perhaps atoms might just correspond to knots in this ether. . . . But this idea soon fell out of favor, and when relativity theory was introduced in 1905 it . . . treated space as just some kind of abstract background, with no real structure of its own. But in 1915 general relativity introduced the idea that space could actually have a varying non-Euclidean geometry—and that this could represent gravity. . . . There were nevertheless immediate thoughts that perhaps at least electromagnetism could be like gravity and just arise from features of space. And in 1918 Hermann Weyl suggested that this could happen through local variations of scale or "gauge" in space, while in the 1920s Theodor Kaluza and Oskar Klein suggested that it could be associated with a fifth spacetime dimension of invisibly small extent.

STEPHEN WOLFRAM, 2003

The idea of adding another dimension to the matrix, of course, is not new. As far back as the 1850s, Georg Reimann (1826–1866) suggested that "the force that causes light (electromagnetism)" was a "ripple" of this extra dimension.[41] This idea of a Reimann space, which is linked to the realm of hyperspace, influenced one of Einstein's associates, Hermann Weyl (1885–1955), a mathematician who studied under David Hilbert (Minkowski's compadre) at the University of Gottingen. Seeking Einstein's goal of uncovering, in Einstein's words, one underlying "uniform field, which would see gravitational and electromagnetic fields as different . . . manifestations of the same field,"[42] Weyl expanded upon Faraday's and Maxwell's work on electromagnetic forces that turn in on themselves. He "had the idea of extending Einstein's theory to incorporate electromagnetism, so that Maxwell's equations would also acquire geometrical significance. Weyl's idea was to introduce a scale, or gauge, that varied from point to point and whose variation round a closed path in spacetime would encapsulate the electromagnetic force."[43]

This work was expanded by Theodor Kaluza (1885–1954), a mathematician from the University of Könisberg, who, in April of 1919, wrote Albert Einstein about his idea of adding a fifth dimension to Einstein's four-dimensional spacetime model. His goal was to unify theories of light and gravity. This was the holy grail for Einstein, the Grand Unification Theory, which was never achieved. When his ideas were combined with those of Oskar Klein (1894–1977), the Kaluza-Klein extra-dimensional model for the universe was born. "Oskar [*sic*]

Klein proposed that Weyl's gauge should be viewed as a phase and that spacetime should be viewed as having a fifth dimension consisting of a very small circle. Mathematically, Weyl's gauge variable gets multiplied by i (the square root of –1) and is periodic."[44]

This theory created a hidden "internal" space, internal to normal space, where electromagnetism and the general theory of relativity could join forces by sharing an underlying gauge invariant, or uniform scale geometric symmetry, which could give rise to both the three forces that had already been combined and gravity.

Kaluza saw this fourth spatial dimension as circular. Where ten quantities were needed to describe a point in 3-D space, Kaluza realized that fifteen quantities would be needed to describe a point in 4-D space. Since four of these extra five quantities could be used to describe electromagnetism, Kaluza hypothesized that some other arrangement of these extra five, along with the original ten quantities, might also explain gravity.[45]

"A five-dimensional cylinder would never [have] dawned on me," Einstein wrote Kaluza. "At first glance, I like your idea enormously." However, because this concept was purely a mathematical one, with no direct relation to physical reality, Einstein abandoned the idea.[46]

Since electromagnetism was "an artifact of this fifth dimension of space-time," and since "electric charge is quantized," Dietrick Thomsen (1984) asked whether or not gravity could be quantized as well via a Weylian Kaluza-Klein model. This concept has become fashionable in attempts to help verify the modern-day superstring theory, which sees all particles and fundamental force fields as vibrations of minuscule supersymmetric strings. Finstad (2003) states that the "internal" space of this extra dimension is "small." Possibly, the gateway to this realm is through the tight corkscrew geometry of the electron, whose charge is a fundamental constant.

In a November 2007 interview, theoretical physicist Lisa Randall put it this way: "The primary puzzle is why gravity is so weak. The theory we're exploring is that gravity may be concentrated someplace else, in an extra dimension. The region where it is concentrated may be a sort of parallel universe, with completely different chemistry and forces.[47]

> Since the nature of the cosmos recedes into infinite orders of infinitesimals or infinites on all sides, there is an ever-present component of ineffability, which denotes the unconstructability of the infinite. This ineffability is a hallmark of life itself, with its characteristic ever-newness.[48]

TWISTING FASTER THAN LIGHT SPEED

The Chaldeans took the number 120 as basic. It is 1 × 2 × 3 × 4 × 5, and is divisible by 6, 12, and 24. This explains how we have 360 degrees in a circle and the numbers 60 and 24 in our measurement of time. But seven is a maverick number. It will not divide exactly into any number and it seems to be connected with "squaring the circle" for the ratio of the diameter to the circumference is almost exactly 22/7. The belief that 7 is a sacred number is traceable to Chaldean sources and early development of arithmetic.

J. G. BENNETT

As mentioned earlier, according to one hypothesis, elementary particles spin at a speed in excess of the speed of light, and this speed, $1.37 \times c$, is an integral feature of the structure of spacetime. Morris notes that superstring theory, as attractive as it is, "has never yielded a testable prediction,[49] but the theory that I am discussing here *is* testable. Were Goudsmit and Uhlenbeck right? The orthorotational speeds of electrons can be measured, and their finding that electrons spin in excess of the speed of light can be verified.

As with any freestanding pendulum or orthorotating object, spinning particles set up their own coordinates in space irrespective of the movement of the Earth. They simultaneously generate, through special combinations (the atoms of the periodic table of elements), the dimension we call the physical world. This idiosyncratic principle intrinsic to orthorotation—which is used in gyroscopes for navigation—has profound implications, for it suggests the need for a stationary ether, what Poincaré was looking for in 1904 in the work of Newton: "an absolute physically preferred reference frame."[50]

As discussed in the first chapter, the number 137 shows up on the macroscopic scale as well, as in the "fine structure" constant, which measures the ratio of matter to energy; it also shows up subatomically in the strength of the electromagnetic force inside of atoms (1/137 is the probability that an electron will absorb a photon).

From the work of Oersted and Maxwell, we could hypothesize that the ether itself may be a matrix of tiny vortices that spin at speeds in excess of the speed of light. This grid would be analogous to the web-like illustrations that so often appear in textbooks, drawn to describe the bending or sinking of space around large bodies (like planets). Maybe this grid of hyperspatial spinning vortices is the warp and woof that gives the ether its stability as well as its complex structure and undetectable nature.

EINSTEIN'S CRITICS

Today's scientists have substituted mathematics for experiments, and they wander off through equation after equation, and eventually build a structure which has no relation to reality.

NIKOLA TESLA

The most important criticism against Einstein's theory of relativity comes from the Nobel Prize committee, which simply refused to honor the theory because it was "highly speculative" that the bending of light around stars "was not a true test of Einstein's theory," and because "there were other ways to explain this phenomenon . . . [and]

Mercury's orbit."[51] To solidify the point, the Nobel committee specifically awarded Einstein for an entirely different discovery, concerning the photoelectric effect.

In his masterwork, *New Model of the Universe,* P. D. Ouspensky (1931/1971) discusses some important ideas concerning the structure of spacetime. He also critiques the roles of both relativity and quantum mechanics as describers of reality. Ouspensky writes that faith need not play a role in establishing the conviction that the world was created "on purpose," as opposed to the view that life and the universe evolved from accident or chaos. Harking back to such philosophers as Pythagoras and Aristotle, Ouspensky says the answer lies in our definition of form.[52]

Ouspensky begins his discussion with the realization that the Euclidean structure of space had to be altered in order to explain the phenomena of light and electromagnetism. Part of the changing concepts of three-dimensional space involved the introduction of a fourth coordinate to account for time. Mathematicians such as "Gauss, Lobachevsky . . . and especially Riemann . . . were already considering the possibility of a new understanding of space."[53] From this premise, Ouspensky, like Pythagoras, Musès, and Wigner, asks what the relationship of physical science is to mathematics. He concludes that the two propositions—1) every mathematical proposition must have a physical equivalent, and 2) every physical phenomenon can be expressed mathematically—have no "foundation whatever, and the acceptance of them as axioms arrests the progress of thought along the very lines where progress is most necessary."[54]

Ouspensky has put his finger on an important point, for on the one hand, mathematics has virtually performed miracles in explaining the physical universe, yet on the other hand, mathematics has been mistaken for reality. The map is not the territory. Musès agrees, yet he also argues that the introduction of imaginary numbers better explains reality. This is a paradoxical situation, which, by its nature, cannot be completely resolved.

Ouspensky clearly points out that quantum mechanics explained all physical phenomena as "phenomena in motion," but that some scientists went so far as to suggest that it might also explain thought and behavior. Ironically, by defining physics as "matter in space and phenomena in matter,"[55] it may very well be that physics and quantum physics could, in fact, actually achieve explanations for psychological and parapsychological phenomena as well. However, if quantum physics seeks eventually to subsume psychology and parapsychology, Ouspensky warns that we should keep in mind that the measures in physics are "artificial."[56] He pointed out that the four constants discovered by Planck—the velocity of light; gravity; energy, divided by temperature; and energy multiplied by time—are actually artificial measures. "The law of gravitation, according to Newton, described the relationships between such things as planets, but did not truly prove the actual existence of a physical phenomena."[57] Ouspensky notes that Newton was well aware that his theory was not concrete proof of an actual gravitational force. However, through the years this basic truth has been lost. Just because the gravitational relationship is explained by $m_1 m_2/d^2$ (where m_1 and m_2 are the mass of two respective objects and d^2 is the distance between them), it does not prove the existence of a force. Problems associated with the issues of action at a distance, unified field theory, an all-pervasive ether, and the fundamental structure of space are involved here.

Like Bergson and Dingle (whose discussions follow), Ouspensky questions Lorentz and Einstein's explanation of the Michelson-Morley experiment as proof that measuring sticks contract in the direction of their motion. Part of their reasoning also included the impossibility of proving the existence of the ether. "The theory of the lengthwise contraction of bodies, deduced not from facts, but from Lorentz' transformations, became the necessary foundation of the theory of relativity."[58]

Ouspensky, however, errs in his discussions of Einstein by stating that his theories are only mathematical and do not correspond to physical reality. The most famous counterargument can be found in Einstein's prediction that light would be bent by gravity, which was supported by measurements taken during an eclipse of the Sun in September 1919. Einstein's biographer R. W. Clark points out that the "star displacement" found by Eddington at this time confirmed Einstein's theories, thereby "mark[ing] a turning point in the life of Einstein and in the history of science."[59]

It should be noted that important questions have been raised about the validity of Eddington's findings. The Royal Academy actually sent two expeditions to cover the eclipse. Eddington went to the island of Príncipe, off the coast of West Africa, while Crommelin and Davidson went to Brazil. It was cloudy in West Africa the day of the eclipse, but the weather was much better in Brazil. Out of sixteen photographic plates, Eddington was able to use only two; even at that, there were only a small number of stars near the sun that could be measured. The physicists had hoped to measure a broad range of stars on all sides of the solar body, but this was not possible. While Eddington regarded his slim measurements as confirming Einstein's theory, the measurements from Brazil actually confirmed Newton's predictions. Ludwik Silberstein, ironically a proponent of relativity, said at the time, "There is a deflection of the light rays, but it does not prove Einstein's theory."[60]

Whether the deflection of starlight by the Sun was caused by the curvature of space (that is, its idiosyncratic geometry around stars), the downward pressure of the ether, or the attraction of light by gravity is subject to debate. In the 1930s, this issue was hotly contested by inventor Nikola Tesla in a series of newspaper articles. "To say that in the presence of large bodies space becomes curved," Tesla remarked, "is equivalent to stating that something can act upon nothing. I for one, refuse to subscribe to such a view."[61] For Tesla, light was seen bending during that solar eclipse because of the Sun's tremendous force field. One way of viewing this is that the ether is warped by large bodies, just like ripples in a stream warp around rocks.

Another way is to assume that light particles, that is, photons, do indeed have *mass*.[62] Einstein has argued that photons have *energy* equivalent to Planck's constant (6.626×10^{-34} joules per second). But energy and mass are equivalent. That would be the simplest explanation, and this would also explain the redshift in receding stars and a blueshift in stars that are headed our way.

Ouspensky was right in asserting that Einstein's theories are separate from quantum physics. However, by taking into account relativity, specifically the inability of mass to move at the velocity of light, Dirac was able to explain the orthorotational properties of the electron and discover antimatter. Dealing with the incorporation of $\sqrt{-1}$ as equivalent to the time component, Ouspensky writes:

> The "special principle of relativity" is supposed to establish the possibility of examining together and on the basis of a general law facts of the general relativity of motion, which appear from the ordinary point of view to be contradictory, or to speak more accurately, the fact that all velocities are relative and at the same time the velocity of light is non-relative, limiting, and maximal. Einstein finds a way out of the difficulty created by all this by:
>
> 1) Understanding time itself, according to the formula of Minkowski, as an imaginary quantity resulting from the relation of the given velocity to the velocity of light.
> 2) Making a whole series of altogether arbitrary assumptions on the border of physics and geometry [such that space is curved around gravitational bodies].
> 3) Replacing direct investigations of physical phenomena . . . by purely mathematical operations.[63]

Ouspensky states that Einstein sought to coordinate mathematics, geometry, and physics into one whole. He questions, however, whether this in fact has been done. In quoting Einstein, Ouspensky writes that although "Einstein regards Minkowski's 'world' as a development of the theory of relativity . . . in reality the special principle of relativity is built on the theory of Minkowski."[64] Einstein's postulates, according to Ouspensky, are really built on Lorentz's transformation and Minkowski's imaginary time unit, and not the reverse. "Because if it were the reverse, then it remains incomprehensible on what basis the principle of relativity is actually built."[65]

Ouspensky continues his astute critique of Einstein's theories by stating that "the velocity of light is violated by the curvature of light in gravitational fields," for if light has no mass, then this would seem impossible, and so Einstein gets out of the problem by "curving the space!"[66] Ouspensky ends the section by quoting Einstein at length (approximately 3,600 words), allowing Einstein (and then Eddington) to expound upon the various inner contradictions in relativity.[67] Ouspensky therefore concludes that relativity creates an unnecessary impediment to physics, whereas quantum physics has a better chance of leading the scientist closer to the truth.

Dingle and Ouspensky and even Einstein himself certainly point to certain inconsistencies in the theory. One of the clearest problems involves the curving of space and yet the virtual denial of the ether. The dismissal of an obvious truth (that is, the Oersted/Faraday/Maxwell force-field etheric grid), and its replacement by a more abstract vacuous medium, is unsatisfactory. In order to expand upon the present laws of physics, a scientist must shake loose his biases in order to create a better picture of reality that not only reconciles the problems inherent in relativity, but also better explains the facts.

Ouspensky points out that Euclidean three-dimensional space did not explain all physical phenomena. There *was* a need to introduce concepts regarding an imaginary time unit. However, he also notes that four-dimensional spacetime also falls short of explaining all phenomena. "The proof of the artificiality of the four-dimensional world in new physics lies first of all in the extreme complexity of its construction, which requires a curved space. It is quite clear that this curvature of space indicates the presence in it of yet another dimension or dimensions."[68] Therefore, Ouspensky concludes, there are not enough dimensions. He also states that mathematics alone cannot describe the true nature of the various dimensions.

Leon Jacobson writes in the translator's preface of Henri Bergson's book on time, *Duration and Simultaneity:* "Our intelligence, looking for fixity, masks the flow of time by conceiving it as a juxtaposition of 'instants' on a line." He says that through intuition, one contacts the "duration" of time. Our senses and consciousness distort "real time" and science ignores the effects of time. Both science and philosophy mistake "the conception of reality for reality itself."[69]

According to Bergson, there is a difference between the *measurement* of time and the *experience* of flowing time or real duration.

1. To measure it, it must be spatialized.
2. "We think of the experienced flow of our inner duration as motion in space; and the next, when we agree to consider the path described by this motion as motion itself." [70]

As an example, one can spend a week in Paris or a week at home and the experience of these two times is very different from a subjective point of view, even though the clock time is precisely the same.

"Time then seems to us 'like the unwinding of a thread like the journey of the mobile (the Earth). . . . But there is the unwinding [time] and our own biological and psychological time—' which are two different things."[71] By breaking down the motion into equal instants, we do not discuss "the real time that endures. Instant–simultaneities are imbedded in flow–simultaneities." But if time "is not finally convertible into experienced duration, it is not raw time, but space, which we are measuring."[72] Then what is time? Paul E. Anderson, reflecting on the teachings of Gurdjieff, put it this way:

Time, the "Merciless Heropass," is meted out for us all. And it is so doled out that we have only limited amounts. Yet we act as if this were not so. . . . Our attitude about Time does not reflect any of its real meaning and real point. . . . We must never, never miss the opportunity of taking advantage of what is before us at any moment . . . because once having lost it, you will not be able to get it back.[73]

Bergson points out that "[t]he motion induced slowings of time allegedly uncovered by Einstein's theory of special relativity is not convertible into duration. These times are not truly lived, but are purely fictional. Thus Einstein's unreality of multiple times betokens the singleness of real time."[74] Bergson distinguishes between real time and spatialization into objects, events, and clock time, or everyday life.[75] He also discusses the illusion of Einstein's slowed clocks as analogous to seeing a man in the distance as a midget. If we were to conclude that he truly was smaller, then this would correspond to doing the same thing. "It is insofar as [clocks] run more slowly and tell a different time that they become phantasmal, like people who have degenerated into midgets."[76]

Introducing Bergson's book was Herbert Dingle, a professor from the University of London, who writes that this analogy was good only up to a point, as subjective experience could be eliminated from the experiment, making the issue: If the clock does in fact slow down, why does it do so? Dingle suggests that Lorentz's theory (which Einstein's is directly derived from) suggests that the "clock is actually slowed down by its motion through the ether,"[77] whereas Einstein's theory postulates no ether, or none that could be detected.

Since mass increases with increase in velocity, this fact must be taken into account whether the increase in mass is due to a gravitational force or to an increased absorption of etheric energy. A clock traveling at almost the speed of light will, according to Einstein, attain close to infinite mass. Such an increase in weight should certainly affect the workings of the clock. Dingle raises further questions regarding the actual viability of Einstein's physical theory. Since according to

Einstein, in a situation of uniform motion one cannot determine which clock is moving and which is at rest, we could just as easily deduce that the moving clock is stationary and the stationary clock is receding at almost the speed of light.

$$\rightarrow \rightarrow \text{ moving clock } \rightarrow \rightarrow$$
$$\leftarrow \leftarrow \text{ stationary clock } \leftarrow \leftarrow$$

Inherent in Einstein's theory of relativity is the interchangeability of the reference system. In the case of the stationary and moving clocks, either one can be slowing down, depending on the point of view. This leads to the "contradiction that each clock could be proved to work both faster and slower than the other."[78] As a practical example, oftentimes while sitting in a train I have seen another train moving nearby. If my train is at rest, then sometimes it seems that the other train is at rest and my train is going in reverse. This is a common illusion, which often occurs to train travelers. It is also an inherent flaw in Einstein's paradigm, because a further extension of the same principle suggests that, depending on a person's point of view, either the Earth is spinning on its axis or the Earth is stationary and the cosmos is spinning about it.

Einstein's most recent biographer, Walter Isaacson, notes that Einstein's 1905 special theory of relativity "only refers to observers moving at constant velocity relative to one another" and that "relativity . . . asserts that the fundamental laws of physics are the same whatever your state of motion."[79] He then refers to a situation whereby a lady is on a plane flying at uniform motion above the Earth: "The woman in the plane could consider herself at rest and the earth as gliding past. There is no experiment that can prove who is right."[80] It seems to me that if the plane were to run out of gasoline, that would be an experiment that would solve the conundrum.

Clearly, there is an external all-encompassing frame of reference, which Einstein neatly ignores in his theory of relativity and corresponding supporting examples involving people, light beams, clocks, and trains, and that is the movement of the Earth around the Sun. This fundamental aspect of reality does indeed set up an overarching reference system, which supersedes these examples because it does allow one to determine which clock is stationary and which is moving near the speed of light. The regulated aspects of the Solar System also sets up a rather firm basis for determining the march of time and the simultaneity of distant events, both of which Einstein insists are relative—but they are not at all relative when taking this third sine qua non point of view with the two he cites (for example, being on the plane or on the ground).

The problem with this thought-problem is the premise. Ultimately, the question is not whether one observer in a plane or another on the Earth can tell if he is stationary or moving. The issue is what is objectively true. Obviously, the plane is flying over the Earth. Isaacson does, of course, observe that this problem applies only to entities moving at uniform motion. Once spin and acceleration enter into the equation, it is much harder to make the case. This is where Einstein's 1919 general theory of relativity comes in, because it deals with acceleration.[81]

It seems that Einstein was creating these problematic thought experiments because of his wish to abandon the ether, some absolute frame of reference; for some odd reason, physicists have continued to embrace this obviously subjective stance.

After publishing his remarks on the internal inconsistency of Einstein's theory in *Nature* and *Science,* Dingle asked the scientific community to respond. The only relevant comment came from Max Born in 1963, who argued that Dingle "advanced the wrong problem." It is Dingle's view that ignoring or explaining away Bergson's argument delivers a critical blow to relativity. He also asserts that the mental dimensions of time have been essentially ignored by the physicists. He concludes that this internal aspect of time cannot be understood by equations.

A FRESH LOOK AT MICHELSON-MORLEY

While the Michelson-Morley experiment established that the ether cannot be detected, it did not disprove the existence of the ether.[82] According to James Coleman, in his *Relativity for the Layman,* there are a number of other possible explanations why the Michelson-Morley experiment failed:

- The Earth may have dragged the ether. Coleman, however, states that this ether drag should be detected when looking at the stars.[83]
- The ether was actually detected, but because its effects were so minimal, they were discarded.
- "The explanation which had the most appeal in accounting for the negative result of the Michaelson-Morley experiment was one that was literally dreamed up for the purpose. It is the so-called Fitzgerald-Lorentz contraction. In 1893, Fitzgerald suggested that all objects contracted in the direction of their motion through the ether: 'The force of the ether contracted them. This would adequately explain the results.'"[84]

One wonders, however, if this means that the Earth is constantly contracting as it too travels through the ether, and, if Lorentz is correct, then the orthorotation of the Earth might generate an important centrifugal component that could offset any direct contractual motion through the ether. Coleman does not get into the "why" of the contraction, and whether the contraction continues with the motion or contracts only initially. In any event, Einstein's special theory confirmed the contraction hypothesis! It stated:

First Postulate: The ether cannot be detected. All motion is relative—only to something else. "A stationary ether would possess absolute motion because it would be the one thing in the universe, which would be motionless. But we can only detect relative motion."[85]

Second Postulate: The velocity of light is always constant relative to an observer.

Third Postulate: "The fundamental laws of physics are the same whatever your state of motion."[86]

"The effect of length contraction can be stated simply: Whenever one observer is moving with respect to another, whether approaching or separating, it appears to both observers that everything about the other has shrunk in the direction of motion." According to Coleman's calculations, a plane traveling 750 miles per hour would shrink to about the diameter of a nucleus.[87]

When time is involved in the description of distance between two points, it appears mathematically as the fourth dimension, t (with x, y, z).

Einstein showed, on the basis of the Special Theory, that while two events might be simultaneous for one observer, they are not necessarily simultaneous for all observers. In fact, it can very easily occur that two "happenings" could be simultaneous for a first observer but for a second observer one event would precede the other. And for a third observer the order of the events could even be reversed![88]

For these kinds of reasons, there are many theoreticians who think that Einstein's theories are incomplete, and perhaps even built on flawed data. Other, more recent critics are Ian McCausland (1999), a physicist from the University of Toronto, and Bjorn Overbye (2007), a physician with a background in physics from Norway.

According to these authors, Michelson thought a "static ether penetrates all objects." To explain the inability of Michelson and Morley to detect differences between the two light beams in their famous experiment, Lorentz and also Fitzgerald thought that the Michelson interferometer that was going through the ether (as opposed to the one that was at right angles to it) shrank the exact minute amount to cause both beams to look like they arrived back at the same time. According to this idea, as suggested by the work of Poincaré, the entire theory of relativity derives from these Lorentz-Fitzgerald contractions; space contracts and time dilates; a moving clock really goes slower than a stationary one; and a plane traveling through space is really shorter (by a minute amount) than an identical plane that is stationary. Einstein came along and said that the ether is "superfluous," yet his theory was derived from other scientists—including Michelson, Lorentz, and Fitzgerald—who assumed the ether was a reality.

Overbye points out that Einstein's theory, which was supposedly original, suggests that the speed of light is a universal constant and that "all movement is relative to other movement, but not to a universal resting ether." Yet this theorizing was really based on the previous publications of scientists like Mach, Poincaré, and Lorentz, which Einstein did not cite. Further, Overbye writes, orthorotation is not relative motion, but absolute.[89]

Another critic at the time was Swiss theoretical physicist Walter Ritz, who said time was irreversible. Einstein, on the other hand, in debates with Ritz "defended Maxwell-Lorentz electromagnetic time symmetry, [and, as a result,] microscopic reversibility remains a cornerstone of our current paradigm."[90] Time, of course, in the humble opinion of this author, is not reversible, and according to Fritzius, who has written about Walter Ritz on the Web, Einstein frankly knew this.

Overbye also introduces the reader to two scientists who *were* able to detect the ether by replicating the Michelson-Morley experiment: George Sagnac and Dayton Miller. Also, Einstein was anticipated on his work on Mercury's perihelion by Paul Gerber, whose equations predated Einstein's by eighteen years; Gerber achieved these results by using another method. In 1911, Ernst Gherkin accused Einstein of plagiarizing Gerber, and Einstein countered by saying that Gerber's equations were in error,[91] but the implication really was that if the same findings could be derived by an alternate method, then Einstein's theory in this regard was unnecessary.[92]

Dayton Miller's ether detection experiment is covered in Ronald Clark's biography of Einstein, in which Clark describes a meeting between Miller and Einstein and quotes a real fear of Einstein's that motivated him to assert that Miller must be wrong in his calculations.[93] To paraphrase Einstein himself, if the ether can be detected, then relativity is wrong. However, none of the other Einstein contemporaries critical of Einstein cited in Overbye's essay is even mentioned in Clark's or Isaacson's major Einstein biographies. Considering that Ritz coauthored an article with Einstein and that Einstein defended himself against accusations that he plagiarized Gerber, the fact that these critics are missing from the most important biographical literature on Einstein supports the contention that the mainstream scientific community is simply unwilling to take on Einstein, even though his theory has serious problems.

Further, where Clark (1971) correctly credits Minkowski for laying the mathematical basis for Einstein's theories on a four-dimensional spacetime continuum, Isaacson (2007) completely miscasts Minkowski's role by devoting little more that four or five sentences to this important contributor and repeating a misconstrued Minkowski comment about Einstein's lack of mathematical skills three times in his book.

Although Isaacson has written a superb, even an inspired biography, sadly, his decision to obfuscate Mikowsky's true role in Einstein's work helps delegate this important mathematician to almost nonperson status. And at the same time, this decision aids in minimizing the role of imaginary numbers in explaining physical phenomena, a topic that is completely missing in this biography. Minkowski has been excised out of many other books on Einstein as well.

Einstein himself goes so far as to suggest that he "owed even more" of his theory of relativity to "the Scottish philosopher David Hume" than he did to Mach or Poincaré! Hume's ideas stemming from the tabula rasa theory of Aristotle point out that "mental constructs . . .

were divorced from purely factual observations."[94] So, Einstein admits to making liberal use of the importance of the "observer" in formulating his paradigm-shifting theory, but in no way does Einstein deal with the *consciousness* of the observer as a force in itself or with the "inner space" where the observer's mind is located.

EINSTEIN'S ETHER LECTURE

One misconception attributed to Einstein is that he believed the ether did not exist. What he said was that it could not be detected. In a letter to Lorentz, Einstein wrote in 1916, "I agree with you that general relativity theory admits an ether hypothesis."[95] And four years later, in an address at the University of Leiden in 1920 entitled "Ether and the Theory of Relativity," Einstein concludes that since "space is endowed with physical qualities, in this sense . . . there exists an ether. According to the general theory of relativity, space without ether is unthinkable, for in such space there is not only no propagation of light, but also no possibility of existence for standards of space and time. . . . [Thus] we may assume the existence of an ether, only we must give up ascribing a definite state of motion to it." It seems, according to Isaacson, Einstein needed to resurrect ether "to explain rotation and inertia."[96]

In his treatise, Einstein draws a parallel between ether, space, and the gravitational field (which he sees as potentially scalar, that is, having a directionless magnitude), on the one hand and matter associated with the electromagnetic field on the other.

In the address, Einstein spells out the two distinct realms that he had been unable to unify, gravitational ether (that is, space) and the electromagnetic field associated with matter. He mentions his meeting with Hermann Weyl, and Weyl's idea of creating another dimension to connect gravity and electromagnetism. However, Einstein felt strongly that this theory, although ingenious, did not conform to reality. As he hints in his Leiden address, electromagnetism may in fact be formed from the complex structure of space.

The following section begins with the generally accepted premise that Einstein entirely dismissed ether theory. Even though, as we have seen, Einstein clearly did not entirely dismiss ether theory, the essential end product of Einstein's theory, for all intents and purposes, was to dismiss the need for ether.

Einstein's problem stemmed from his obvious wish to protect his theory of relativity, which had been heavily influenced by Ernst Mach and Mach's principle, which hypothesized that all matter in the universe is interdependent. Thus, Einstein would write the young mathematician Karl Schwarzschild on January 9, 1916, "Inertia is simply an interaction between masses, not an effect in which 'space' of itself is involved, separate from the observed mass."[97] Schwarzschild, Isaacson points out, disagreed.

Now, four years later, in 1920, after reconsidering the necessity of the ether, for instance, as a means to propagate light, Einstein changed his mind, so that now his view of a rotating body was that it did not just obtain its inertia from, and in relation to, all the rest of the matter in the universe, but also on its own accord, due simply to "its state of rotation" because "space is endowed with physical qualities."[98] The end result for Einstein was that "he admitted outright . . . that he was now abandoning Mach's principle."[99]

Although Einstein's most recent biographer covers this highly significant change of heart, and in fact does so in superb detail, Isaacson couches the entire argument as *an apology* for Einstein, as Isaacson begins this chapter with the premise that now that Einstein was past the age of forty, his creative years were behind him, he had become "blockheaded," and, worse than that, quoting Einstein himself, "the intellect gets crippled."[100] In tracking down Isaacson's endnote, one finds that this is a letter from Einstein to his friend Heinrich Zangger, dated August 11, 1918, and one also finds the cited article by Clive Thompson, "Do Scientists Age Badly?" *Boston Globe*, August 18, 2003. At least Isaacson explored Einstein's 1920 Leiden lecture. Einstein's other most famous biographer,

Roland Clark,[101] on the other hand, leaves it out or overlooks it even though he had created a list of twenty of "Einstein's more famous scientific papers and lectures."[102] Resurrected on the Web, this watershed lecture was, as we have seen by Einstein's two most prominent biographers, dismissed, overlooked, or diminished in importance for the simple reason that the end result of the lecture was to essentially discount some of the most important findings associated with relativity.

However, in his own way, Isaacson suggests a reason why Einstein needed to resurrect the ether, and it was due to his work on the photoelectric effect (which gave him the Nobel Prize) and the new findings of Louis de Broglie and Niels Bohr and their corresponding mathematical equations.

Einstein had said that a photon acted like both a wave and a particle. Influenced by the yin/yang sign, this led Bohr to his principle of complementarity, which simply codified this idea by agreeing with it. And it also led de Broglie to the assumption that elementary particles must also be waves. The wavelength of the particle, such as an electron or a proton, "would be related to Planck's constant divided by the particle's momentum," which would "turn out to be an incredibly tiny wavelength" but a wavelength nonetheless.[103] Said in another way, the wavelength of an elementary particle can be calculated by dividing its quantum of light energy by its mass times velocity: h/mv.

By this time, Bohr had come up with his Nobel Prize–winning idea of a solar system–like model of the atom with the electrons orbiting the nucleus of protons and neutrons in discrete orbits. Since Tesla's ideas will be discussed at length in the following chapter, it is important to note Tesla's historical relationship to the development of this theory. Bohr derived this model from working in England with J. J. Thomson, the discoverer of the electron, and Ernest Rutherford, who, working with Thomson, came up with a model for the atom which became the basis for Bohr's model that led to his Nobel Prize.

The key that is missing from this history of the structure of the atom is Nikola Tesla, who spoke before the Royal Society in England in 1893, a lecture that Thomson attended and which Tesla published in the electrical journals and in the book *The Inventions, Researches and Writings of Nikola Tesla,* which Rutherford was known to have treasured and read.[104] In this lecture and its related one from Columbia University presented a year earlier, which was, in part, influenced by the work of Lord Kelvin, Tesla not only preempted Rutherford and Bohr's atomic model, but he also preempted Einstein's idea that photons could operate like waves or particles, stating as follows:

1. That energy through the ether could be transmitted "as transverse vibrations" or "by independent carriers."
2. "That molecules and their atoms spin and move in orbits in much the same manner as celestial bodies, carrying with them, and probably spinning with them, ether, or in other words, carrying with them static charges."[105]

Not only did Rutherford study Tesla's compendium when he performed his wireless experiments, but also Bohr spoke at the commemoration of Tesla's hundredth birthday in 1956 in Yugoslavia. This was at the same time that "the Institute Electrotechnical Committee agreed to adopt the name 'tesla' as the unit of magnetic flux density," for example, MRIs are measured in teslas.[106]

Einstein, possibly unaware of Tesla's connection, nevertheless was heavily influenced by this new wavelike model for the atom being developed by Bohr and de Broglie. Rather than simply view the atom precisely like a solar system with particle-like electrons circling the nucleus in prescribed orbits, this more sophisticated model realized that the electron was also a wave, or a discrete pulse that had a particular energy. Thus, the quantum leap, which occurs when electrons jump from one orbit to another, is not a magical event, because the electron is really a wave that shifts its position or point of focus, and this can occur when electrons emit or absorb photons.

Thus, Einstein resurrected the necessity for the ether because the wavelike model for the elementary particles was clearly becoming a new gateway to a further understanding of the fundamental structure of matter and space and one needed an ether as a means to transmit waves through space. However, Einstein was now also caught in a clear contradiction. He had stated that the ether by its nature could not be detected and further, that if it could, relativity was wrong. And yet, by 1920, he reassessed his view and now believed that the inertia and momentum of a spinning object or particle was evidence that, in his own words, "space had physical qualities."[107] The implication here is that the very fact that matter has inertia is evidence of the presence of ether.

The next chapter explores an intriguing avenue for achieving Einstein's dream of creating a unified field theory, by uniting gravity with the ultimate structure of matter.

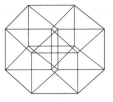

5

ETHER THEORY REVISITED

Einstein says that there is no ether and at the same time, practically he proves its existence. For example, let's consider the speed of . . . light. Einstein states that the velocity of light does not depend on the rate of movement of the light source. It's correct. But this principle can exist only when the light source is in [a] certain physical environment (ether), which cuts down velocity of light due to its properties. Ether's substance cuts down the velocity of light in the same way as air substance cuts down acoustic speed. If the ether did not exist, then velocity of light would strongly depend on the rate of movement of the light source.

SHAPKIN/TESLA? 2002

The World Wide Web has generated a new and unprecedented forum for theoreticians. Before the web, articles and books could certainly be self-published, but in general, serious treatises had to go through a chain of command before seeing the light of day. This had the positive effect of preventing many flaky ideas from infiltrating the mainstream. But on the negative side, important theories that threatened the prevailing worldview—such as those regarding the ether—were blocked as well.

From the web I discovered Anaxagoras, an ancient Greek theoretician who speculated that "the ether was the most subtle substance in creation: the mother of all phenomena," and "that atoms were vortexes in the ether, a theory picked up 2500 years later by Lord Kelvin."[1] By the 1800s, ether theory was well accepted, but, as we have seen, it was discredited in the next century, and essentially stays that way at the beginning of this century as well. Nevertheless, reinterest in ether theory is emerging. Metaphysician David Wilcox put it this way: "This ether is a source of *tremendous* energy that is in constant motion, flowing in and out of all objects in the universe—just as a candle flame is constantly absorbing new wax and oxygen and radiating new heat and light, but still continues to exist as a measurable unit." Wilcox also tells us, "[P]hysicists John Wheeler and Richard Feynman have calculated that the energy in the volume of vacuum space *contained within a single light bulb* is concentrated enough to bring *all the world's oceans to the boiling point*!"[2]

In a controversial paper published on the web, purported but not proved to be Nikola Tesla's actual words, M. Shapkin introduces some intriguing ideas about ether theory and the criticism of some of Einstein's findings.

According to this paper, Tesla solved the Michelson-Morley dilemma with two reasons why the ether was not detected. The first was that "the ether is electrically neutral, [and thus] it very poorly interacts with the material world."[3] The second was simply because the Earth was moving *with* the ether in its path around the sun.

In the first instance, according to Wilcox, who was interpreting the work of theoretical physicist Harold Puthoff, the ether is essentially neutral because "energy is applied equally in all directions, and thus, to us, it has no typically measurable movement or force—it cancels itself out . . . usually to zero."[4] This is the so-called zero point, which supposedly is a reservoir of virtually unlimited amounts of energy, although it is difficult to access.

In Shapkin's paper, Tesla's second reason why the ether was not detected is supported by an example of a boat caught in a whirlpool. If a person on the boat were to place his hand in the water to feel its motion in relation to the boat, there would be no difference detected because the boat and the water are moving as one. Similarly, the Earth is circling the Sun within, and accompanied by, the ether.

According to this Teslaic view, while the ether is difficult to detect electronically, it is, in fact, quite easy to detect mechanically. Tesla claimed to have accessed the ether through mechanical action, rather than through just electrical means.

As Tesla understood it, electromagnetism produces longitudinal or soundlike waves in the ether, which are misinterpreted when conceived of as just transverse waves through the air, for instance, like those used by radio broadcasting systems. Where radio evolved using only the transverse aspect through the air (even though broadcasting stations also had a ground connection), Tesla made more use of the longitudinal aspect, which was through the ground. Where radio broadcasts waves were longer, Tesla reduced the length to create "very short waves." This pulse resembles sound waves and those produced through mechanical action (in an analogous way, as by a jackhammer). Since Tesla claimed he knew the resonant frequency of the Earth, that is the channel he

Fig. 5.1. A whirlpool in the North Atlantic

tapped in to to use as his carrier wave. "The transmitter," Tesla wrote, "will generate vibrations in synchronization with the period vibrations of the earth."[5] According to Tesla, moving at the speed of light, this pulse would not emanate in all directions, but rather would "travel in definite straight line paths determined beforehand along the surface of the globe with [little or] no dissipation or energy."[6] "Invariably, these waves, [when reduced to just a few meters in length, would] just like the atmosphere, follow the curvature of the earth and bend around obstacles. . . . Ultrashort waves also have the same property."[7]

In his September 11, 1932, article published in the *New York Herald Tribune,* Tesla criticizes Einstein's view that space is curved. "Nothing could be further from my mind," Tesla writes. "Space cannot be curved," because space, by definition, can't have properties. "This downward deflection always occurs irrespective of wavelength," Tesla said. Further, it would be "all the more pronounced the bigger the planet." As far as I know, other than this controversial Shapkin article, Tesla's theory of gravity has never been published. The downward deflection of light and other electromagnetic waves would be caused by a force field, which Tesla most likely saw as the absorption of ether by stars and planetary bodies. If the electromagnetic waves are short or very short, they would be more susceptible to the effect of the planet absorbing ether. Thus, the waves would be less likely to disperse, and that is why they would travel in essentially straight lines. And conversely, if the waves were longer, as used in radio broadcasts, the transmissions would scatter more widely. Further, it was Tesla's contention that ether was present in a vacuum, and that was why his illuminating tubes gave off a corona when the charge leaked into the air. Tesla writes about it in later life:

In 1896 . . . I brought out a new form of vacuum tube capable of being charged . . . [to] about 4,000,000 volts. I produced cathodic and other rays of transcending intensity. The effects [in the vacuum], according to my view, were due to minute particles of matter carrying enormous charges. . . . Subsequently, those particles were called electrons. One of the first striking observations made with my tubes was that a purplish glow for several feet around the end of the tube was formed, and I readily ascertained that it was due to the escape of the charges of the particles as soon as they passed out into the air. . . . [Since] it was only in a nearly perfect vacuum . . . the coronal discharge [that leaked out] proved that there must be a medium besides air in the space, composed of particles immeasurably smaller than those of air, as otherwise such a discharge would not be possible. On further investigation, I found that this gas was so light that a volume equal to that of the earth would weigh only about one-twentieth of a pound."[8]

Three years later, "in 1899, I obtained mathematical and experimental proofs that the sun and other [similar] heavenly bodies . . . emit rays of great energy, which consists of inconceivably small particles animated by velocities vastly exceeding that of light."[9]

According to Tesla, these are cosmic rays, which penetrate the Earth, coming from the Sun in an almost instantaneous fashion. But at the same time, on their way to the Earth, they collide with "cosmic dust," thereby causing a secondary radiation that showers the Earth, not just from the Sun, but from all the other stars as well.

If we consider Tesla's vacuum tubes (and also any ordinary incandescent lightbulb), if all the air has been removed, what is left? Tesla says it is ether, an extremely light gaseous substance. Whether or not this ether exists, what certainly exists inside vacuum tubes is the intersecting light from every star. The ether is at least that, so there may be two distinct levels to ether:

1. A fundamental medium pervading all of space that is constantly absorbed by all elementary particles so they can maintain their spin.
2. The intersecting light rays from every star (and galaxy).

Light cannot be anything else but a longitudinal disturbance in the ether, involving alternate compressions and rarefactions. In other words, light can be nothing else but [something analogous to] a sound wave in the ether . . . . [Thus] light propagates with the same velocity irrespective of the character of the source. Such constancy of velocity can only be explained by assuming that it is dependent solely on the physical properties of the medium, especially density and elastic force.[10]

In an interview Tesla gave to the well-known columnist Joseph Alsop Jr., Tesla reveals something he had tried to keep to himself before it was ready, because it was his plan to harness cosmic rays to generate electrical power. These he saw as extremely short rays of great penetrative abilities coming from the Sun that could travel as much as fifty times the speed of light. "'I have detected,' Tesla said, 'certain motions in the medium that fills space, and measured the effects of these motions.' The results of the experiments had led him 'inescapably' to the conclusion that such bodies as the sun are taking on mass much more rapidly than they are dissipating it by the dissipation of energy in heat and light."[11]

Although a gaseous medium, Teslaic ether is not ethereal at all, but rather is highly dense, so much so that matter appears airy in relationship to the ether! (Shapkin/Tesla?) According to Tesla, "This gas was so light that a volume equal to that of the earth would weight only about one-twentieth of a pound," yet the "elasticity to density" ratio for this universal gas would be "800 trillion times greater than air." The end result was the existence of a medium that restricted the speed of electromagnetic pulses to 186,000/second, or the speed of light.[12]

Any form of acceleration (such as a g-force) is really a manifestation of the ether impressing itself upon the material body. "The ether tries to return itself to the initial state by compressing our world, but intrinsic electric charge within material world substance obstruct[s] this."[13]

Gravity, from this point of view, is not a *pulling* force, caused by an attraction of large planetary bodies, but rather is caused by the ether itself *pushing* down on the Earth as it makes its way through the cosmos. What we call gravity, according to Shapkin/Tesla?, is the "universal compression" of ether. Its presence and great density is what limits the speed of light. According to this etheric view, light speed is actually *slowed down* by ether because of ether's great density. He also suggested that some cosmic rays travel not just fifty times the speed of light, but also thousands of times faster than the speed of light, implying that there are higher octaves for light speed. (This will be further explored in chapter 13.)

This results in a radical view regarding matter and energy: "All energy is drawn from the environment," said Tesla. "There is no energy in matter except that absorbed by the medium."[14] This can be compared with the more traditional view as espoused by Fritjof Capra in his landmark text *The Tao of Physics:*

> The striking new feature of quantum electrodynamics arises from the combination of two concepts; that of the electromagnetic field, and that of photons as the particle manifestations of electromagnetic waves . . . [with] each type of particle corresponding to a different field. In these "quantum field theories," the classical contrast between the solid particles and the space surrounding them is completely overcome. The quantum field is seen as the fundamental physical entity; a continuous medium, which is present everywhere in space. Particles are merely local condensations of the field.[15]

CURLING LIGHT

If light is considered to be a wave, then it must pass its vibration through a medium (that is, the ether). However, if, as Planck and Einstein discovered, light is also a quantum of energy, it need not pass through a medium, and so would move through empty space. This model described

light and other forms of electromagnetic energy as being made up of photons: wave-packets that had the properties of both waves and particles.
With the advent of this theory explaining how light could travel through
empty space, the concept of the all-pervading ether faded away and science began to look at space as comprising a vacuum.

However, in the 1930s and forties, Nobel Prize–winner Paul
Adrian Dirac resurrected the ether, yet did it in a way that somehow
did not raise the ire of the academic elite, who had neatly done away
with it. Dirac postulated that empty space, the vacuum, was not really
empty, but instead "filled with electrons in negative energy states."[16]
Providence College physicist E. Gora continues, "Space is thus like
the 3-dimensional ocean in 4-dimensional space-time. What is below
this surface (antimatter) is not normally observable."[17] Professor Gora
also points out that "space is never quite empty. Light is partially
absorbed, primarily by random atoms and particles."[18] Echoing Weyl's
and Kaluza/Klein's idea of creating an extra dimension involving a
small round space of precise size, which they call gauge invariance, I
think Dirac's idea of a sea of antimatter is compatible. Parapsychologist
Andrija Puharich bumps up the complexity of this extra dimension by
linking it to different mental dimensions as well, but it is still based on
something analogous to this Kaluza/Klein fourth-dimensional small
round space. Puharich tells us that an extension of Dirac's original conceptualization implies infinite negative energy states, equivalent opposites to the myriad positive ones.[19]

Fritjof Capra writes in his watershed book *The Tao of Physics,*
"Einstein's theory . . . says that 3-dimensional space is actually curved,
and that the curvature is caused by the gravitational field of massive
bodies. Wherever there is a massive object, e.g., a star or a planet, the
space around it is curved and the degree of curvature depends on the
mass of the object."[20] Capra goes on to point out that the gravitational
field *is* the curved space; it does not curve the space. "In general relativity, the gravitational field and the structure or geometry of space are
identical."[21]

When [Einstein] first began formulating general relativity . . . he had . . . figured that the bending of light by the gravitational field next to the sun would approximate . . . what would be predicted by Newton's theory when light was treated as if it were a particle. But now [after a number of years of working on calculations, circa 1915], using his newly revised theory, Einstein calculated that the bending of light by gravity would be twice as great, because of the effect produced by curvature of space-time.[22]

As stated above, this hypothesis was countered by Tesla, who said flat out that Einstein was wrong, and that the ether must exist. In an article that appeared in the *New York Herald* (9/11/1932), Tesla wrote that light bends around stars and gravitational bodies because of a force field. Professor Gora agreed that Tesla might simply be defining the phenomena of light bending around large bodies in a different way that ultimately did not contradict Einstein's findings. Is "space" curved, or is it some type of etheric field "warped" by some other effect, which might still correspond to Einstein's equations? In any case, Gora went on, "[O]ne should keep in mind that the Einsteinian view is not the curvature of ordinary three-dimensional space, but of four-dimensional space-time."[23]

NEW ETHER THEORISTS

Tesla's views can be further developed by combining the work on gravity and ether theory done by Price and Gibson (1999), Ed Hatch (2003), Vencislav Bujic (2001), Ron Heath (2003), Warren York (2003), and David Wilcox (2000). The ether is "omnidirectional" and "the very basis of space-time," says York.[24] The photon, he goes on, may be the ether itself, "but in a different form."[25] The ether gives rise to all forms of matter. In this view the elementary particles (such as the electron) are seen as standing waveforms in the ether spinning much like whirlpool eddies caught in a running stream (this is not very different from the fundamental vibrating strings of superstring theory). Ether from the cosmos is constantly supplying centripetal action to the electron's high orthorotational speed in the elementary particle (which I suggest is 1.37c). Much like the action of a centrifuge, which separates particles by weight, or specific gravity, the process of spin generates differences within the particle, which results in a polarity, its direction of spin determining its pole. This also causes field discharges that transfer the incoming energy. We know from classical physics that the spin of an electron, by its very action, generates a magnetic field. Where electric fields are caused by the separation of positive from negative, magnetic fields are caused by electrical fields in motion. Strong fields, such as found in permanent magnets, are caused when many charged atoms are aligned in the same direction. And the velocity of propagation between these charged particles is the speed of light.

The conclusions derived from these recent theories and some of my own thinking can be expressed as three postulates:

Postulate 1: *The syncopated pulsations, that is, the polar reversals, of the cosmos generate the fundamental spin of the elementary particles.*

MIT professor Ed Fredkin speculates that the universe either is or operates like a computer, and is thus based on a binary system—or series of on/off switches. Complexity would derive from an algorithm, which he defines as "a fixed procedure for converting input with output, for taking one body of information and turning it into another." By adding in a feedback loop, which is called a recursive algorithm, a "simple programming rule [can produce] immense complexity."[26]

Certainly, polarity reversals are a rhythmic fundamental inherent in the construction of the cosmos. Where the planet Earth reverses its poles about every 700,000 years, the Sun, at a lightning pace, flips poles every eleven years. Human brain waves oscillate at a much more rapid pace, that is, from one to sixty-plus cycles per second. "I believe that in order to make any real progress, one would again have to find a general principle wrested from Nature. . . ."[27] "[And this method] justifies us

Fig. 5.2. Nikola Tesla in his laboratory in 1898 sending 500,000 volts through his body to light the fluorescent tube that he is whipping around in a circle. The photo was taken with a strobe by his photographer Dickenson Alley.

in believing that nature is the realization of the simplest conceivable mathematical ideas."[28]

In the early 1880s, Nikola Tesla discovered a rotary effect obtained from polar reversals. He called the effect "the rotating magnetic field." He used a naturally flowing alternating current—that is, a current that was reversing its poles and its direction of flow thousands of times per second—to efficiently generate and distribute electrical power and to create a spinning motor. He did this by harnessing two (or more) currents out of phase with each other.

Tesla's discovery came after an intense illness, probably a severe flu, that, in the process of almost killing him, caused him to have supersensitive abilities. For instance, he claimed he could hear a fly land on a table and that he could detect objects in the dark with his forehead. Many believe that his crown chakra was awakened.

Pygmalion seeing his stature come to life could not have been more deeply moved. A thousand secrets of nature, which I might have stumbled upon accidentally I would have given for that one, which I had wrestled from her against all odds and at the peril of my existence.[29]

An armature attracted to the North pole will spin, thus producing the rotating magnetic field.

N 0	N N	0 N	S N	S 0	S S
↖	↑	↗	→	↘	↓
0 S	S S	S 0	S N	0 N	N N
1.	2.	3.	4.	5.	6.

Fig. 5.3. Each quadrant (such as N + 0 and 0 + S, shown below it) represents two circuits catty-corner to each other. The North-South pole in quadrant 1 (top left corner to bottom right corner) starts to reverse itself and appears in 2 as North-South, in 3 as 0/0 (because it is reversing) and then as South-North in 4, 5, and 6. The other circuit (top right corner to bottom left corner) demonstrates the same pattern as it reverses, but it does so out of phase with the first circuit. The arrow, representing an armature, is attracted to the North pole. Note that North moves in a circular pattern. This is how a motor works. The whirring of the motor, or its spin, is caused by these two out-of-phase, catty-corner alternating circuits. Spend a moment to reflect on this diagram.

On August 23, 1893, Tesla displayed this principle before a thousand electrical engineers at the Chicago World's Fair specifically in relationship to planetary motion. Present were such notables as Sir William Preece and Silvanus Thompson from Great Britain; Elihu Thomson, one of the founders of General Electric; Elisha Gray, inventor of one of the first fax machines; and the "honorary chairman," Hermann von Helmholtz, probably the premier scientist of the day.

Tesla set up a large platform "energized" with his rotating magnetic field. Inside, he placed several large and small brass balls. When the alternating current turned on, "the large balls remained in the center and the small ones revolved around [them]." Vacuum bulbs were lit throughout

Fig. 5.4. This illustration, which clearly resembles a spiral galaxy, apparently also has plus and minus charges. One of Tesla's logos that he often placed on his stationary, the drawing certainly has cosmological implication linking, perhaps, the birth of the universe to the inventor's discovery of the rotating magnetic field.

the Electrical Hall, "but the demonstration, which most impressed the audience was the simultaneous operation of numerous balls, pivoted disks [illuminated like crystals] and other devices placed in all sorts of positions at considerable distances from the rotating field."[30]

If the universe is constructed on the same principle—that is, polar reversals of out-of-phase opposite energies—this pulsation would initiate the primary spin responsible for, on the macroscopic scale, the spinning galaxies, stars, and planets, and, on the microcosmic scale, the spin of the elementary particles and the structure of atoms. On the Earth, this fundamental spinning force would result in energies associated with the jet stream and weather patterns such as hurricanes, tornadoes, and lightning storms. This would also suggest that this omnipresent etheric field is an AC current, possibly reversing its charge billions if not trillions of times per second.

Bujic (2001) writes that "[e]ther can flow in a variety of forms: parallel, rotational [and] converging, or propagation can be in the form of longitudinal pulses."[31] This last idea is associated specifically with Tesla's view of electromagnetic energy being transmitted either transversely or longitudinally through the ether "as alternating compressions and expansions similar to those produced by sound waves in the air."[32]

Ron Hatch, a director of NavCom Technologies (navigation communication) along with his brother Ed, uses ether to explain the four physical forces in the universe. We will start with his analysis of the first three:

1. Electromagnetism, caused by the two fundamental properties of ether: compression and shear or twist oscillations. *Compressive* oscillations cause the *electric* component; *shear* or *twist* oscillations, the *magnetic* component. "It is the external ether distortions caused by the particle's rotating standing waves that produce rotating compressive . . . and shear oscillations" and corresponding electric and magnetic fields. Ether, then, would constantly be absorbed by elementary particles, which would, in turn, as they keep their spin, generate this energy back out into the field as electromagnetic emanations.

2. Merging *group-spin* effect, when all the protons spinning and aligned together in the nucleus, create the strong nuclear force. This would be a synergestic effect, and that is what is holding together the nucleus of the atom.

3. *Localized shear* fields for coupling protons to electrons create the neutron, for the weak nuclear force. Each elementary particle is a spinning top with poles, so electrons have a plus side and a minus side, as do protons. Direction of spin determines which way the poles run. A neutron is a combination of a proton and an electron. It differs from a hydrogen atom because it is more compact and is inside the nucleus. Most likely, *directional spin effects,* which the Hatches call "localized shear fields," augment either a normal attraction between an electron and a proton as found in the hydrogen atom or a super strong attraction when they combine in the weak nuclear force to create a neutron. In that case, the negative pole of an already negative electron is attracted to the positive side of an already positive proton.

According to Hatch (2003), the electron's spin is also determined by the compression and shear or twist oscillations of ether. The shells of an atom (the regions in which electrons may be found) are caused when the rate of expansion of the "sphere" (the group-spin of the nucleus) begins to approach the speed of light. At that moment, a shock wave ensues, causing an orbital shell for an electron. The frequency and diameter of the nucleus (the number of protons and neutrons from the classical point of view) determine the different number of orbital shells. "All matter is nothing more than encapsulated energy in the form of standing-wave structures, with all interactions taking place at the speed of light. This is the absolute reaction time allowed by the reacting medium."[33]

Postulate 2: *Order presupposes conscious design.*

The idea that spin can be created by out-of-phase alternating currents or polar shifts offers a different rendition of what is called the big bang theory. The primary force responsible for the creation of a spinning universe and all its offshoots (such as spinning galaxies) could start with the One and its reflection. A circle or zero is, paradoxically, one thing. One and its negation, yin and yang, start the ball rolling and create the initial alternating current spin. This primary ignition operates lawfully, thus

Fig. 5.5. Two images of the yin/yang symbol. In the first, we can see the seeds of each pole present in the other. In the second, the idea of spirals inside spirals is evident.

principles of consciousness (e.g., its orderly structure) are inherent in its design (see chapter 1 for the definition of *consciousness*).

The key here is to reevaluate what we mean by the term *conscious*. It is a complex word that involves some form of primary awareness, demonstrated by electrons that repel other electrons and are attracted to protons. The initial yin/yang has an inherent order and it is this built-in antagonistic order that generates the asymmetric motive force of the universe.

Wilcox suggests that every level of the universe is an outcropping of a "unified ether of spacetime energy that vibrates multi-dimensionally according to specific numerically-based harmonic principles of light, sound, and geometry. . . . There is a tremendous energy that is in constant motion flowing in and out of all objects in the universe."[34] Polarity reversals, rhythm, spin, motion, inertia, angular momentum, and the principle of organization should be added as other fundamental components. All stem from the first postulate, represented by the yin/yang symbol arranged in a hierarchical order: a primary syncopated antagonism creates spin, galaxies, fundamental particles, and so on.

This multitiered arrangement of the ether may be set up like a rainbow or a harmonic scale. If this were the case, white, which comprises all the colors, would correspond to the All, or first-tier ether, and the seven levels (Ray of Creation) would correspond to what Gurdjieff and Ouspensky (1949) call the Law of Octaves or Laws of Three and Seven. This is seen in the harmonic scale as do, re, me, fa, sol, la, ti. According to this schema, the two half-tones between me-fa and ti-do are interruptions in the vibrations that cause changes in direction. These two half-tones, the second more powerful than the first, are the flies in the ointment that allow for variety to take place, such as evolution. It is interesting to square this view with that of a traditional physicist, George Gamow, who wrote:

In 1925, a French physicist, Louis de Broglie, published a paper in which he gave a quite unexpected interpretation of Bohr quantum orbits [the electron shells around the nucleus of an atom]. According to de Broglie, the motion of each electron is governed by some mysterious *pilot waves*. . . . Assuming that the length of these pilot waves is inversely proportional to the electron's velocity, de Broglie could show that various quantum orbits . . . were those that could accommodate an integral number of pilot waves. Thus, the model of an atom began to look like some kind of musical instrument with a basic tone (the innermost orbit with the lowest energy) and various overtones (outlying orbits with higher energy). One year after their publication, de Broglie's ideas were extended and brought into more exact mathematical form by the Austrian physicist Erwin Schrödinger whose [Nobel Prize–winning] theory became known as "Wave Mechanics."[35]

Postulate 3: *Matter requires ether to maintain itself, ultimately to sustain the spin and standing waveforms of its elementary particles.*

This process, which is continual, transforms ether into what we call electromagnetic fields. For instance, the Earth is constantly absorbing a tremendous amount of ether to keep all its elementary particles spinning. This is what we call gravity. On the macroscopic scale, this constant influx of ether produces the Earth's North and South poles. On the microscopic scale, it produces the various EM fields around the particles that make up the atoms, elements, molecules, and so on.

"The absorption [of ether] is proportional to the density of the medium."[36] Thus, what we experience as gravity on the Earth is really an ether gradient flowing *into* the planet equally from all sides simultaneously. This process is amplified because of the great requirements for ether demanded by the molten core. In this way, perhaps the Earth could be considered a small star with a crust (mantle) around it. The planet therefore acts like a funnel that is drawing in ether continually. This is what gravity is.

This etheric view may be at odds with the fundamental implications of what is known as Mach's principle, a hypothesis first supported by physicist and philosopher Ernst Mach: "Matter there governs inertia here."[37] According to this principle, all stars and planets are connected through some supergravitational force. Where Mach's view unifies the entire universe, the etheric theory of gravity seems to nullify any need for hypothesizing some overarching link among all bodies in the universe. This is to say that, in the Newtonian sense, there would be no action at a distance. What looks like gravity might really be competition for ether. However, matter might still be linked throughout the cosmos by some other resonance model, scalar potential, or nonlocal effect.

Bujic offers an explanation for the fourth physical force of the universe: "Gravity is when *ether* accelerates. Inertia is when *matter* accelerates."[38] Warren York puts it another way: "If you move aether through mass you get gravity. If you move mass through aether, you get inertia."[39] These end up as essentially the same thing, though their starting circumstances are different. In the first instance, the object (say, for example, an airplane at rest) is absorbing ether simply because of its mass. In the second instance, the object is accelerating (the airplane taking off), and so is actively changing its inertia by

Fig. 5.6. Artist's rendition of gravity

moving through the ether at an increased velocity (in this instance, the velocity of the Earth plus the speed of the plane); it is thereby absorbing more ether.

Compare this to Einstein's own view, pointed out by Isaacson in his new Einstein biography. Newton has found that "inertial mass always equals gravitational mass."[40] Einstein saw these two instances as "equivalent. . . . The effects we ascribe to gravity and the effects we ascribe to acceleration are produced by one and the same structure."[41] But what is this structure? Einstein spent a lifetime trying to understand precisely what gravity is. Bujic's ether theory explains precisely what this "same structure" is.

Thus, in the etheric view, what we call gravity is actually the never-ending absorption of ether by elementary particles to maintain their spin and thus their integrity. And when the body accelerates, the absorption rate increases accordingly. That is what a G-force is.

In fact, Einstein himself provides a hint that the ether has density when he writes in his 1920 Leiden lecture paper that transverse waves cannot be produced in a fluid; they can be produced only in a solid:

When in the first half of the nineteenth century the far-reaching similarity was revealed, which subsists between the properties of light and those of elastic waves in ponderable bodies, the ether hypothesis found fresh support. It appeared beyond question that light must be interpreted as a vibratory process in an elastic, inert medium filling up universal space. It also seemed to be a necessary consequence of the fact that light is capable of polarization that this medium, the ether, must be of the nature of a solid body, because transverse waves are not possible in a fluid, but only in a solid. Thus, the physicists were bound to arrive at the theory of the "quasi-rigid" luminiferous ether, the parts of which can carry out no movements relative to one another except the small movements of deformation, which correspond to light waves.[42]

THE HIGGS FIELD AND DARK MATTER

Refusing to accept ether has led physicists into some awkward positions. While ether theory offers a credible explanation of gravity, numerous articles echo the simple fact that physicists have yet to integrate gravity into their modern-day spacetime model. Einstein spent the latter half of his life looking for a unified field theory, and he never succeeded in finding one. In articles that appeared in 2001 in such prestigious forums as *Science News* and the *University of Chicago Magazine,* it seems clear that nineteenth-century ether theory has reappeared in the guise of the theories of Peter W. Higgs, a physics professor from Edinburgh University: "To patch these flaws in the standard model, theorists proposed the existence of some sort of influence that permeates all of space, weighing down particles passing through it. This cosmic molasses is called the Higgs field . . . a pervasive field in the universe that . . . could bestow mass on all fundamental particles that have mass."[43]

Using sophisticated particle accelerators and spending billions of dollars, physicists around the world continue to look for the "Higgs boson" that makes up the field. Some, the author Weiss tells us, have actually dubbed the Higgs boson the "God particle." Yet after tens of thousands of collisions, the great binding entity has yet to be found. Might this just be an expensive way to deny the existence of the ether while trying to explain its effects?

The recent discussions about dark matter may amount to the same thing. According to Marcia Bartusiak, physicist and author for *Discover,* and Meg Urry, chair of the astronomy department at Yale University, we can only see about 10 percent of the matter that exists in the universe.

> We assumed that this expansion [from the big bang, 13.7 billion years ago] should be slowing down due to the pull of gravity. In 1998, however, two teams of astrophysicists discovered that the expansion is actually speeding up. They observed a mysterious form of "energy" that opposes gravity and is causing the galaxies throughout

the universe to move apart faster and faster. . . . The discovery of "Dark Energy" is arguably the most important scientific break-through of the last fifty years.[44]

In fact, it may be the very ether itself, a form of fundamental or pre-energy, corresponding to what these theorists suggest may be a universal scalar field, the realm that gives rise to elementary particles and feeds their spin. After completely rebuking ether theory, Bartusiak—216 pages into her treatise—makes a reassessment that maybe the ether exists after all. She writes, "Could even more dark matter, perhaps another type of matter altogether, be strewn uniformly through the cosmos, marking a return of the ancient Greek ether, but recast in a new form?"[45]

For arbitrary or semantic reasons, the missing mass has been called "dark matter" or "dark energy." In fact, this so-called dark energy may really be "light energy." Each point in space contains the intersecting light from every star: hundreds of billions of stars and galaxies and hundreds of millions of years of radiating starlight create a long-standing photonic grid that certainly permeates every corner of the universe at the speed of light. The ether is thus at least a sea of photons in some type of grid of light; what we call a photon is most likely a pulse in the medium that has, as Bohr has suggested, both wavelike and particle-like properties.

Without even referring to the more esoteric "etheric substance," this invisible omnipresent photonic grid of starlight may account for a large portion of the so-called missing energy that the astronomers are looking for. When astronomers calculate the amount of energy radiating from a galaxy, do they take the time component—which can be 100,000 light-years or more per galaxy—into account? Further, it is simply a misnomer to say that outer space is empty; it is jam-packed with energy. As an example, if a solar panel was placed anywhere in the universe, it could convert star or galaxy light into electrical energy. It is self-evident that invisible electromagnetic energy permeates space.

A question that could be asked here is whether the actual light

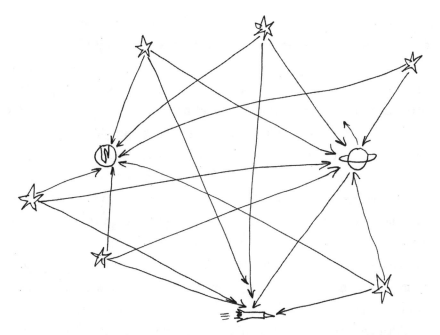

Fig. 5.7. Star grid. There is no place in the cosmos where the light from all the stars is not present.

from a distant star has physically traveled the many miles to reach the Earth at the speed of light or has an idiosyncratic pulse emanated from that star; has this vibration (or wave packet) been transmitted from sector to sector through the ether to reach the Earth? Interestingly, just as the light from every star is present one way or another right here on our planet, the reverse is also true, namely, the emanations from our planet and Sun have also, throughout the millennia, radiated out to every reachable point in space. Thus, just as "they" are here, we are there. This is fundamental: everything is interconnected through this complex photonic grid.

GRAVITY WAVES

In the mid-1990s, Tom Van Flandern caused a stir in the physics community when he was able to get a paper published in some of the physics journals calling into question the traditional answer to the problem of the speed of gravity propagation. The physics community had, for the most part, shut down most debate over theoretical problems inherent in Einstein's theories; as a rule, it became difficult for physicists to question potential inconsistencies in the theory of relativity.

Van Flandern created an interesting thought problem. If the Sun was sending out a gravity wave to the Earth at the speed of light, it would take about eight or nine minutes to reach the planet, depending on whether the planet was close or far away in its orbit. During that time, however, the Earth could, in theory, not be in gravitational communication with the Sun, and thus would move farther away. The only way around this, for Van Flandern, was to adopt Newton's assumption that gravitational interaction is essentially instantaneous. Gravity waves would therefore travel at many times the speed of light.

Although this view generated some controversy, it ignores a fundamental aspect of the Sun/Earth connection, namely that the Sun has never been out of contact with the Earth as, most likely, the Earth and other planets were born from the Sun, and spun out from it in the early universe when the solar system was created. Thus, the premise was false.

Ether theory, on the other hand, coupled with Ouspensky's interdependent view, looks at the structure of the solar system in a holistic way. The solar system could be seen as one entity, so to speak, so that the Sun is not so much continually attracting the Earth, but rather they are moving together as a unit through the galaxy. Ether would be flowing through both of them, but what looks like the Sun "attracting" the Earth is really the flow of ether through a small body as compared to a vast body. The pressure from the inflow of ether would be offset by:

1. The Earth's rotation, which would convert certain aspects of the incoming gradient into its magnetic field, such as in the Van Allen belts, jet stream, and so on.

2. Another large-scale galactic etheric movement through which the entire solar system is moving, analogous to Tesla's boat caught in a whirlpool.

Price and Gibson suggest using the word *nether* to mean dynamic or moving ether and *ether* to stand for static ether. In fact, there may be a hierarchy of ethers associated with the solar system, galaxy, and universe. These would be linked to the angular momentum of each system, as well as to a vast relatively static ether, which would lie outside of the local eddies created by each separate hierarchical galactic system.

SCALAR WAVES

The concept of scalar waves may also offer insight into the nature of ether. A scalar quantity involves a directionless magnitude, as compared to a vector, which takes direction into account. Associated with the idea of an all-pervasive gravitational field, it is a ubiquitous quantity of compressed energy, known in some circles as zero-point energy. One wonders if there is a link here to the concept of the "now" or the time component, for time also, this ever-present "now," is also a "directionless magnitude."

The expert on the topic of scalar waves is Tom Bearden, who suggests that the scalar component, missing from modern understanding of electromagnetic (EM) waves, is an important key to understanding the true structure of electromagnetic waves, the link between photons, ether, gravity, Mach's principle, the Higgs field, nonlocality, superstring theory, and Kaluza-Klein hyperspatial models concerning the fundamental structure of spacetime. It is the source of a heretofore unrecognized yet essentially unlimited reservoir of energy.

Bearden writes in *New Energy News,* July 1998, that "whenever an EM wave starts to form, both the transverse and longitudinal waves start to form. However, the transverse wave has a function, which cancels the longitudinal wave." This normal EM wave, now missing the longitudinal component, is "comprised of photons."

Bearden then writes, "A photon is a piece of angular momentum. So it's a piece of energy welded to a piece of time, with no seam in the middle, so to speak. What this 'piece of energy' represents . . . is a dynamic oscillation of the energy density of 3-space." If the EM wave can be created in such a way as not to block the longitudinal aspect, it would leave a "longitudinal EM wave" that has a scalar potential and therefore can interact with gravity and thus time, making it able to operate in nonlocal fields.

For Bearden, classical physics has chosen to leave out discussions of scalar waves, which he defines as "any static stationary ordering in the virtual particle flux of a vacuum." Scalar waves are created when the electromagnetic energy of a particle is enfolded into itself. "Scalar waves pass through the electron shells of atoms and interact with the nucleus. They are continually absorbed and emitted by all nuclei in the universe." Bearden states that the Sun constantly emits scalar waves that penetrate to the Earth's core and help keep it molten. In turn, the Earth radiates back scalar waves to both the Sun and Moon, and all are connected through a "scalar coupling system." In fact, he sees all stars as being interconnected through a scalar coupling system.

Bearden's theories on scalar waves are difficult to understand. Nevertheless, they are worthy of serious study. They are intimately related to Tesla's theories on ether theory and the topic of the longitudinal aspect of the EM wave, to Einstein's thoughts on the link between the scalar component and gravity, to zero-point energy, and to the concept of transcending light speed.

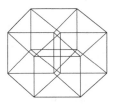

6

ETHER THEORY: THE MENTAL ASPECT

If we look at the evolution of scientific thought, we find that significant advances occurred when individuals began to study supposed anomalies such as the strange paths the planets take as they cross the sky through the seasons. In fact, quite a few secrets of nature were revealed once the Earth's true relationship to the solar system was realized. After Kepler discovered that the planets moved in elliptical paths around the Sun, and the law of planetary motion, Newton was able to formulate the law of gravitational attraction, and Einstein the theory of relativity.[1]

When Galileo's use of the telescope led him to discover some of the moons of Jupiter, Olaf Roemer (1675), a Danish astronomer, was able to utilize their eclipses not only to prove that light had finite speed, but also to calculate that speed with amazing accuracy.[2]

As we have seen, one of the thorns in the side of science was the necessity for discovering the medium that gravity and light waves pass through. Unfortunately, it could not be detected. A number of clever attempts were made to discover this ether, the most notable being the 1880s Michelson-Morley experiment discussed earlier. The failure of this experiment helped nail shut the coffin on trying to locate the ether. The mystery of the structure of the space between the stars has been glossed over ever since.

Although Newton boldly asserted that all material bodies attract one another, he himself did not say whether the gravitational force passed through a material or a nonmaterial medium; still, he left little doubt that he believed the ether exists.[3] Perhaps the nineteenth-century physicist Ernst Mach came closest to explaining the enigma of the fundamental structure of space by going beyond Newton to say that the mass of any material body such as the Earth is dependent upon some type of supergravitational force from all the stars.[4] Whereas Newton believed that motions could be intrinsically absolute, Mach, influenced by Leibniz and by Buddhist thought, said that all effects in the universe were related to all others. This unconventional statement greatly influenced Einstein's theory of relativity, and Einstein wrote Mach to say as much.

Mach was proposing an interconnectedness of all things (stars) and, further, that each part of the system was a microcosm, which in turn reflected the macrocosm. Written in a different form, this was in essence Leibniz's monad theory. If we assume that all things, stars, galaxies, and so on, are interconnected, questions still remain as to the structure and composition of this interconnectedness.

To understand the enigma of the ether, we need to synthesize the historical perspective with modern knowledge. This new data combines parapsychological advancements with the technological achievement of holography. We can give more than lip service to the statement that the microcosm reflects the macrocosm, as we now have a physical apparatus that can duplicate this metaphysical axiom.

HOLOGRAPHY

Holography was developed by Dennis Gabor in the 1960s. To create a hologram, laser light is aimed at a beam splitter, a half-silvered mirror that causes half the light to go through it and the other half to bounce off it. The beam that goes through the splitter is called the reference beam. This beam never sees the object. It bounces off a second mirror (top right in fig. 6.1) before hitting the photosensitive plate (bottom right).

The other beam, the object beam, bounces off the beam splitter and hits yet another mirror (bottom left), which causes it to bounce up to the object, then the light from that beam bounces down to hit the holographic plate (bottom right). The two beams collide as they hit the plate simultaneously. One beam is pure and the other carries the information about the object. The two beams, like intersecting waves, create an interference pattern on the photosensitive plate; when it is processed, the hologram is created.

A laser is needed to view a hologram created in this way. If a laser is pointed at the holographic plate, inside the plate, appearing as if

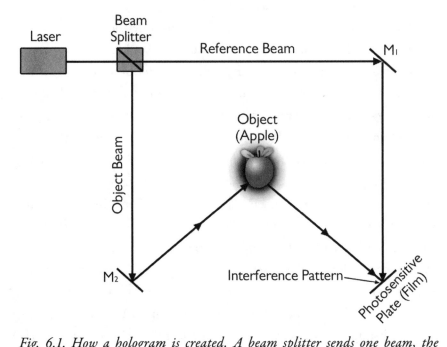

Fig. 6.1. How a hologram is created. A beam splitter sends one beam, the reference beam, off Mirror 1, and another beam, the object beam, off Mirror 2. This second beam hits the object and then collides with the reference beam to create the interference pattern that makes up the hologram.

behind it, a three-dimensional representation of the object is created. By moving to the left or right of the plate, different sides of the object can be seen. It is a true three-dimensional hologram. White-light holograms, which are constructed a little bit differently, can be seen in natural light, and on sci-fi book covers and credit cards.

A holographic image has certain similarities to a regular photograph. In the creation of both a holographic image and a traditional photograph, light from a source bounces off an object, then hits photosensitive film inside a camera. However, in a regular photograph, there is no reference beam and no interference pattern. It is the addition of these two factors and the use of laser light that create the hologram, which behaves very differently. Once the holographic negative is created, it can be cut into a hundred pieces; if a laser shines through any one of them, an image of the entire object will be created. Somehow, the whole is distributed throughout the parts.[5]

The technological advance from two-dimensional to three-dimensional photography has profound philosophical implications because this concrete process changes our very notion of the structure of those three dimensions. This offers a clear prototype of the Leibniz monad, in which each monad reflects every other and each is a microcosm of the macrocosm. This holographic concept whereby the part contains information about the whole is not new to nature. We know from cloning experiments that each cell of a living system contains information, in the form of DNA, for the entire organism. Breaking a mirror into a hundred pieces will produce a similar effect. It is the same principle seen in Kirlian photography: a part of a leaf can be cut off, but when high-frequency electricity is zapped through the leaf, the missing part can still be seen: the so-called phantom leaf effect.

When the Russians announced this finding in the early 1970s, I thought it might be a hoax. How could a leaf have a piece cut off and yet the whole leaf still somehow be present? However, this effect began to be duplicated in laboratories around the world, and I have personally seen the phantom created in my own classroom. In the spring of

*Fig. 6.2. The phantom
leaf effect*

1978, an engineering student brought in a fresh leaf and Kirlian equipment. He cut off a piece of the and zapped it with high-frequency electricity; we then saw a corona of the entire leaf, including the missing part! It is believed that memory and other neurological functions may also be coded in the brain via a holographic-like process. We know, for instance, if a person loses a limb, he or she may still feel it. This is the phantom limb effect.

The inception of holography not only created a new technology; it also changed our views on the properties and structure of the universe. If the brain process called consciousness is created on holographic principles, might it then have within it a code for the monumental unifying energy that created it? This implies, as theoretical physicist David Bohm, states, "a new notion of order. . . . This order is not to be understood as an arrangement of events [in a series]. Rather a total order is contained in some implicit sense in each region of time and space."[6] The possibility that the universe may be constructed on holographic properties alters radically our conception of the world.

THE GALAXY IS A HOLARCHY

Dane Rudhyar, the philosopher and astrologer, has coined the term *holarchy,* which he defines as "a hierarchical structure where each level in the hierarchy is a whole, and all wholes interpenetrate each other." Each "lesser whole" is part of the "greater whole."

We can add a holographic aspect to the definition: Each stratum of the hierarchy can code (in some derivative sense) for every other strata. Different levels are separated by threshold barriers, which have, respectively, their own sets of laws. Yet the principle "As above, so below" still applies. For instance, the structure of an atom has many similarities to the structure of the solar system. Wilhelm Reich, in his book *Cosmic Superimposition,* pointed out the startling similarities between the structure of hurricanes and that of spiral galaxies. Their bird's-eye views look identical. If this is so, what does that tell us about the eye of a galaxy? We know that in the case of the hurricane, the eye is a serene place.

> The Milky Way Galaxy . . . is rotating like a hurricane . . . at an unthinkable 490,000 mph. Meanwhile, the entire galaxy is rushing toward the neighboring Andromeda Galaxy at about 180,000 mph. And just to add some extra drama, our hurtling planet is spinning like a supersonic top, 1,000 mph at the equator.[7]

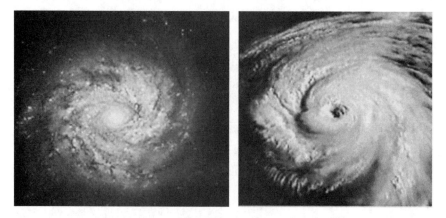

Fig. 6.3. Spiral galaxy (left) and Hurricane Katrina in August 2005. This comparison between galaxies and hurricanes was first explored by Wilhelm Reich in his book Cosmic Superimposition (1955). *The galaxy is spinning approximately 2,400 times faster than the hurricane.*

The spin of the electron is a fundamental constant. Maybe the spin of the galaxy is a higher-order manifestation of this same process. Essential principles like symmetry, rhythm, polarity, movement, angular momentum, spin, form, aesthetics, and synchronicity seem to prevail at microcosmic and macrocosmic dimensions. Such a total order existing throughout the galaxy would have mental as well as physical attributes. Reich's finding of the link between hurricanes and galaxies supports the anthropic, principle the cosmological coincidences, and global design patterns linking the macrocosm to the microcosm discussed in chapter 1. Clearly, in this Wilhelm Reichian instance, the macrocosm is reflected in the microcosm.

New discoveries of incomprehensibly distant galaxies, quasars, black holes, and white holes have bewildered humans. Rudhyar asks whether we have really accepted the knowledge of the vastness and complexity of the cosmos: "New facts demand a new dimension and interpretation of reality." He says we need a holistic approach to solve the enigmas that are being revealed to us. Just as we realized that the Earth revolved around the Sun, we must now realize that the Sun is only one star of billions in the galaxy.

Rudhyar looks at human growth from a comprehensive "evolution of mind" perspective. A newborn begins with an inner connection to the universe. After a time, as Piaget noted, the infant begins to realize that his mother is a separate entity. At this juncture, the baby begins to create an inner symbolic representation of the mother. This is the beginning of memory. As the child grows, he separates himself from the world and becomes self-centered, eventually becoming more altruistic or remaining essentially egocentric.

Similarly, from an anthropologic point of view, humans through the centuries became egoistic by thinking their little world was the center of the cosmos. It took a tremendous change in mentality around five hundred years ago to eventually place the Sun at the center of the solar system. As is well known, Galileo paid the price for espousing this heliocentric cosmology by spending the last ten years of his life incarcerated. According to church doctrine, man, and thus the Earth, was the center of the world, and it was heresy to state otherwise. Before Galileo, Copernicus, the so-called father of the heliocentric view, had to soften his findings that it was really the Sun that was in the center or he, too, would have been castigated.

Rudhyar points out with the title of his book, *The Sun Is Also a Star,* that although we have come to accept the Sun as the center of our world, we must take the next step and realize that the Sun is only one of many suns. In this way, the human comes full circle. (I guess in some holographic way, this is true, in that the center or source can be found inside each of us, but that is a different story!) The Sun is one of many revolving around a larger center, the Milky Way, and who knows what the Milky Way revolves around. One way or another, now that we have come to these realizations, we must learn to transcend our little selfish solar system view and become again like the infant, at one with the universe.

> The essential factor in this transformation of man's consciousness is the transmutation of the "Solar I" into the "Galactic WE." In this "WE-consciousness" the principle of interpenetration operates. . . . Everything not only is related to everything else, but . . . every entity, every mind interpenetrates every other entity. . . . As the consciousness of an individual begins to operate in the spiritual dimension, [man enters] the level of spiritual mind or supermind.[8]

The humbling thought occurs that we have no way of knowing what the galaxy looks like because of the vastness of its size. Yet the mere realization that we *are* in a galaxy extends our consciousness out into the galaxy in proportion to this realization.

Rudhyar separates the ether into three hierarchical interpenetrating realms:

- Biospheric space—the space within which living organisms enter into relationships with each other

- Heliocosmic space—the space within which planets and other material entities relate to each other
- Galactic space—the space within which stars are related to one another

A component Rudhyar does not really address is the time factor. Our solar system sits toward one end of the long spiral of the Milky Way. Even the closest stars can be 10,000 or more light-years away, and the farther stars are more than 50,000 light-years from us. We do not see them as they exist today. We see these stars as they existed when *Homo sapiens* was still a primitive species scrambling against the Neanderthals for control of terra firma. The galaxy is tens of thousands of years older than how we see it today. Other galaxies, seen through telescopes, are millions of years older than we see them today. Unless we can transcend light speed, which is represented graphically by the physical dimension that we so readily see each night in the starry firmament, how could we ever really begin to take our place among our galactic neighbors?

PRE-GEOMETRY AND MINKOWSKI SPACETIME

The diagram on page 134 depicts a "world point" of any event along a "world line." The timelike physical boundaries of the event lie within the light cone. The Observer dot in the center stands for the present. The edges of the cone are "called light-like and describe the history of photons."[9]

Any event occurring outside of the light cone occurs in a realm that is spacelike. By definition, any corresponding particles would also exceed the velocity of light. So far, there is no physical evidence for this realm, although, as noted earlier, Gerald Feinberg coined the term "tachyons" to correspond to the hypothetical faster-than-light energy packets associated with this realm.[10]

The nonlocal EPR (Einstein, Podolsky, Rosen, 1934) paradox, which suggests instantaneous information transfer, can also be explained by operations lying outside of Minkowski's light cone. Quantum physicists

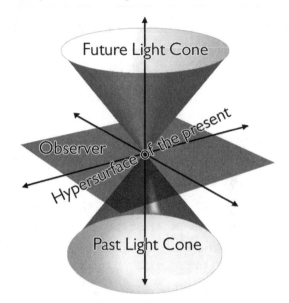

Fig. 6.4. Minkowski's world-point diagram. All physical events occur within the light cone, where the Observer point is the present. Any event falling outside of the light cone would operate, by definition, in a spacelike dimension exceeding the speed of light.[11]

are currently considering the theoretical feasibility of this process. Sarfatti suggests that information and consciousness exist in a quantum physical pre-geometry. He goes so far as to suggest that "psychic information . . . on the level of pre-geometry . . . creates space, time and matter."[12] Just as there are quantum jumps of electrons from one orbital shell to another, information can also "jump" instantaneously. This, for Sarfatti, could correspond to the EPR nonlocal effect on the physical level and to ESP on the mental level.[13]

If we view the world as one totality, using the atom as an analog, then quantum jumping, or instantaneous information transfer, can be accounted for. Each human mind, in this sense, would correspond to different electrons orbiting in their own idiosyncratic shells, circling some primal center; all psyches are part of a single universal consciousness. We are all connected at deeper levels. Thus, a thought that exists in one person's mind could potentially "jump" into the mind of another.

Influenced by Mach's principle (matter there governs inertia here), Whitrow can write that the nature of time is also associated "with the bulk distribution of matter in the universe."[14] Since the universe is assumed to be expanding, this suggests that cosmic time may also be altering. Not only might this affect tachyonic realms and our consciousness, the gravitational constant between elementary particles may also be varying,[15] although there seems to be no evidence that universal constants are being altered.

TIME AND GRAVITY

One of the boldest attempts to refute our current notion of gravity was proposed by P. D. Ouspensky, a unique genius, mathematician, and mystical researcher. Ouspensky points out that Newton's law of gravitation is descriptive of a relationship between two bodies (planets) but that this relationship is not necessarily a mysterious force of attraction.

> Both [Newton] and . . . Leibniz, definitely gave warning against attempts to see in Newton's law the solution of the problem of action through empty space, and regarded this law as a formula of calculation. Nevertheless the tremendous achievements of physics and astronomy attained through the application of Newton's law caused scientists to forget this warning, and the opinion was gradually established that Newton had discovered the force of attraction.[16]

Ouspensky's explanation for the motion of the planets stems from a radically different view of the structure of the universe. Conceiving of the dynamic solar system as a gestalt pattern moving through space, Ouspensky writes:

> If we wish to represent graphically the paths of this motion, we shall represent the path of the Sun as a line, the path of the Earth

as a spiral winding round this line, and the path of the moon as a spiral winding around the spiral of the Earth. . . .

The Sun, the moon, the stars, which we see, are cross-sections of spirals, which we do not see. These cross-sections do not fall out of the spirals because of the same principle by reason of which the cross-section of an apple cannot fall out of the apple.[17]

Ouspensky's idea would replace the gravitational law of attraction by an altogether different concept related to a principle of symmetry, that is, the "movement from the center along radii."[18] For him, the Earth is no more attracted to the Sun than the left eye of an animal is attracted to the right. This is a unique hypothesis derived from a more Pythagorean/Aristotelian view.

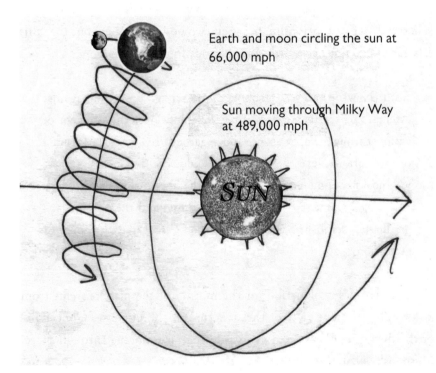

Fig. 6.5. Ouspensky suggests that the Earth and Moon are not gravitationally attracted to the Sun, but rather that they travel together through the galaxy as intertwining spirals.

Ouspensky's corkscrew model of intertwining spirals visually resembles, in stunning form, the movement of electricity through a coil. The magnetic field around the coil resembles the motion of the Moon circling the Earth as they barrel around the Sun. The structure of DNA is also similar. This tendency toward spirality is no doubt a fundamental property of the universe, which would manifest itself geometrically on subatomic levels (such as the spin of an electron as it circles a nucleus), atomic levels (such as in the spiral structure of DNA), and on macroscopic levels (such as in the orthorotation of the Earth, solar system, and galaxies).

A comparable, though very different, model to explain the relationship between the Earth and the Sun has been proposed by Nobel Prize–winning Norwegian astrophysicist Hannes Alfven (1970), who suggests that "giant currents through space from the Sun through the planets, along magnetic lines of force . . . [which he calls plasma] . . . actually transfer angular momentum to the planets."[19]

One potential support for the idea that the angular momentum of our spiral galaxy and solar system causes the Earth to spin can be found in a simple experiment with a magnetized pin, a cork, and a glass

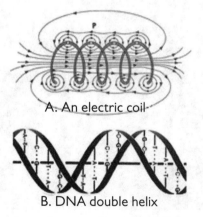

A. An electric coil

B. DNA double helix

Fig. 6.6. Note the similarity between the pattern of electricity moving through an electric coil and the movement of the Earth, Moon, and Sun traversing the galaxy. As the electric current moves from left to right, a magnetic field is created around the coil. DNA also has a spiral structure.

of water. Due to what is known as the Coriolis force of the Earth, the cork will spin clockwise above the equator and counterclockwise below the equator. Water whirlpooling down a drain will follow the same path.

How could the galaxy not have a direct effect on the fate of the Earth? The Coriolis force of the galaxy would generate spin in solar systems and in components thereof. Assuming the galaxy to be one entity, spinning in unison, we can see that the innermost sector (the star clusters near the center) would spin at a rate slower than that of the outermost sector. One way or another, either through the *pull* of gravity or the *push* of ether, the end result will create a whirlpool

Fig. 6.7. The angular momentum of the galaxies may indeed be the ultimate impetus that keeps the planets in their orbits. Note also, through inference, the invisible gelatinous medium (spacetime grid) in which the galaxies evidently float, in this composite photograph.

effect, echoing the structure of the galaxy, seen on the macroscopic scale as the solar system, and here, on the Earth, as its daily spin. This culmination of factors, which results in a Coriolis force on the Earth, whereby the atmosphere spins at faster rates around the equator and at a slower rate at the poles, helps generate our weather patterns and also presents prototypes or analogs that can be applied to subatomic etheric structures. On the microscopic scale, we have the spin of the elementary particles and the electron orbits around the nucleus of atoms; on the macroscopic scale, we have hurricanes, tornadoes, waterspouts, and whirlpools; and on the cosmic scale, we have the spin of galaxies and, most likely, the universe.

ETHER THEORY AND URI GELLER

Why is [psychic] metal bending important? Simply because we do not understand it. . . . [There may be] some new process . . . involved here, which cannot be accounted for or explained in terms of present known laws of physics.

DAVID BOHM AND JOHN HASTED

In 1974, Uri Geller went to the physics department at Birbeck College, University of London, to perform experiments before David Bohm, a theoretical physicist who had worked under both J. Robert Oppenheimer and Albert Einstein, and several other associates including John Hasted, head of the department; Edward Bastin, a professor of language from Cambridge University; and Brendan O'Regan, a writer from Palo Alto. Geller took a key and placed it between his thumb and forefinger. "The bends in the Yale keys occurred gradually . . . [after] 'gentle stroking' . . . sometimes continuing after stroking had ceased."[20] This phenomenon has continued to occur in Mr. Geller's presence for over thirty-five years.[21]

If the ether theory discussed in chapter 5 is correct—namely, that all physical objects require a constant input of ether to maintain their standing waveforms—then what we call gravity is really the pressing down or absorption of ether due to the requirement of the Earth (and its inhabitants) to maintain its mass. Ether theory also offers the possibility of putting forth a new theory of psychokinesis, based on interference with the ether. For example, in the case of Uri Geller bending metal objects with his mind, it may be that his mind is able to screen out ether, that is, prevent the spoon or key from obtaining the ether that it needs to maintain its integrity. For example, suppose Geller's mind field interferes with the influx of ether into the elementary particles that make up the spoon or key. If he somehow stops or slows down the spin of some of the elementary particles, then that object will start to deteriorate, and thus bend.

The very act of thinking involves the redistribution of atoms, specifically the transference of mental information via mRNA (messenger RNA) in the neurons to protein chains at the ends of the dendrites where the new memories are held.[22] Thus, the more a person uses his mind, the more protein in his dendrites and the more complex his brain. According to this view, "all thought is psychokinetic" because *the very act of thinking, by definition, involves a mental event being changed into a physical one:* a thought becomes a memory, that is, mind over matter.[23] When seen in this light, psychokinetic (PK) phenomena are everyday events. It's only when the mind influences something exterior to the body that it's called psychokinesis, but if thought is some type of energy, that is, some type of vibration, then a logical framework for the modus operandi of PK phenomena has been set up. We even know which atoms or elements are being used because mRNA is made up of adenine, uricil, guanine, and cytocine, which are, in turn, made up of carbon, hydrogen, nitrogen, oxygen, and phosphorus. These are the specific atoms and molecules that lie on the interface and act as the liaison between mind and matter.

Circa 1980, I saw Geller bend a large soup spoon in front of a big crowd at a New Age conference in New Jersey. As he rubbed it vigorously with his first finger, the spoon bent *upward,* against what we call

Fig. 6.8. Uri Geller with the Boy Scouts, Reading, England, 1998. From Uri Geller: Mystic or Magician? *by Jonathan Margolis. Printed with permission of the author. Photo credit: Deborah Collinson.*

the force of gravity. At a different time, Senator Claiborne Pell witnessed Uri Geller bending a metal object *upward*.[24]

According to this theory, certain hyperfields generated by the mind (Geller's mind) may be able to block the ether or interfere with what we presently call the gravitational force of an object, making it possible for movements that we see as being against gravity (including levitation) to take place. Obviously, the ramifications of rethinking this process, which we may have mistakenly been calling gravity, are enormous.

THE MIND OF THE OBSERVER

Einstein emphasized the role of the "observer" in his discussion of relativity. However, as a physicist, Einstein had never really intended to integrate the realm of energy of the *mind* of the observer into his physics. He was merely seeking to explain the "physical" world more precisely than Newton. Nevertheless, the role of the observer cropped up and has done so again and again in regard to relativity and quantum mechanics.

The greatest flaw in the efforts of most of these great physicists to use relativity theory and quantum mechanics to explain all phenomena is that they never attempted to include the process of thought, imagination, thought vibration, mental energy, or willpower—that is, mind in the broadest sense—into their cosmological models.

Heisenberg's principle of uncertainty, whereby one cannot accurately determine both the velocity and the position of the electron simultaneously, was formulated because the observer had to disturb the system in order to measure it. From this hypothesis came two major conclusions:

1. The fundamental structure of matter has an uncertain base.
2. The observer can not be separated from the structure of matter.

Heisenberg's formulation had staggering implications concerning the world God created, because it gave matter fuzzy edges and also created a philosophical base for reintegrating the mind into the structure of the physical world.

It is well known that Einstein never accepted the principle of uncertainty, rebutting it with his famous line "God doesn't play dice." The reason why the speed and location of an electron cannot be determined simultaneously has to do with the size and structure of said electron. The only way to measure it is to use some type of measuring device, which inevitably involves photons. Since electrons can absorb and emit photons, once the electron is measured, it is disturbed. However, considering how far we have already come in understanding atomic structure, it seems shortsighted to assume that a process will never be uncovered that will make moot this measuring issue. In other words, who's to say that some future inventor will not be able to conceive a way to measure the movement of electrons accurately without disturbing them?

Ed Fredkin agrees with Einstein on this point. Since Fredkin sees the world as some form of sophisticated computer, his view of its fundamental basis is deterministic. "Quantum physics," Fredkin notes, "at its core, [sees] reality . . . as inherently unpredictable." But this does not make Heisenberg right with his view of this random underbelly. "Any seeming indeterminancy in the subatomic world reflects our ignorance of the determining principles, not their absence."[25] It would seem, however, that Fredkin is in agreement with Heisenberg's assumption that the observer is an important component to a full understanding of the fundamental structure of the universe, because Fredkin sees the ultimate substrate that predates matter and energy as information. In this view, the elementary particles that make up matter and energy are information packets more than "physical" things.

From this view, the use of imaginary numbers to explain physical phenomena, such as by Minkowski[26] and Dirac,[27] makes sense, and it supports the notion that the mind cannot be separated from the structure of matter. Thus, irrational and imaginary numbers are needed to explain higher-order descriptions of physical reality. Minkowski and Dirac utilized the square root of negative one ($\sqrt{-1} = i$) to explain the structure of spacetime. Minkowski created a mathematical equality between the three physical space coordinates and the imaginary unit for one-dimensional time on the macroscopic scale for Einstein's theory of relativity, and Dirac used the same procedure to explain the quantum physical spacetime motion of the spinning electron in a way that did not violate relativity.[28]

Simply stated, the introduction of an imaginary unit *that exists in the mind* facilitated the explanation of the motion of physical phenomena. "Without it, all electromagnetic wave theory and quantum theory could not have been written. Thus the comparatively simple expedient of [adding] one new kind of number more than doubled the entire mathematical power of all previous centuries."[29] Musès concludes, "[T]he factor of consciousness has entered fundamentally into the most recent physics."[30]

THE SEARCH FOR REAL TIME

Before Einstein's theory of relativity, the measurement of distances between two points did not take into account the time component at all. Originally, geometry dealt with distances along a single *x*-axis. Trigonometry and the Pythagorean theorem were used to measure distances between two points for two-dimensional and three-dimensional objects; however "these branches of mathematics could not cope with the additional factor of time, and so an entirely new branch of mathematics, called Tensor calculus, had to be developed."[31] The velocity of light was also integrated into the scheme. By dividing time into mathematical units, Einstein "spatialized time" in order to measure it. Bergson claimed, however, that the "spatializing of time" is not the same thing as "real time." Bergson therefore introduced the idea that "duration," and a corresponding experiential consciousness, must be included in any model that accounts for real time.

> To tell the truth, it is impossible to distinguish between the duration, however short it may be, that separates two instances and a memory that connects them, because duration is essentially a continuation of what no longer exists and what does exist. This is real time, perceived and lived. . . . Duration therefore implies consciousness, and we place consciousness in the heart of things for the very reason that we credit them with a time that endures. However, the time that endures is not measurable, whether we think of it as within us or imagine it outside of us.[32]

Interestingly enough, Roger Penrose, in his book *Shadows of the Mind: A Search for the Missing Science of Consciousness,* goes one step further in seemingly criticizing Einstein's idea of time, yet at the same time completely ignores Bergson's observation. Penrose writes, "[A]ccording to general relativity . . . there is nothing in the physicists' spacetime descriptions that singles out 'time' as something that 'flows'! . . . It is only consciousness that seems to need time to flow."[33]

In an odd way, Penrose is suggesting that if humans did not exist, there would be no "time." Frankly, this seems absurd. Time flows whether or not humans exist. The universe is simply alive with movement, action, and change, which has nothing whatever to do with human consciousness and everything to do with things that happen in time, things that have a past, present, and future.

At the same time, Penrose ignores so many scientists that have devoted their lives to studying consciousness. Bergson is not the only key thinker missing from Penrose's important book. He also ignores Freud and Jung, and even Gurdjieff. My point is simple. If the physicists of such stature as Penrose truly want to fill in the missing links for creating a "science of consciousness," then the work of the giants in the field must be considered. (See *Inward Journey* for a complete discussion on the structure of the psyche and the thinking of these theoreticians.) In some strange way, Penrose dismisses entirely the idea of consciousness and the realm of the unconscious, while at the same time stating that time would not exist if it weren't for consciousness.

Part of the problem with this discussion is that time, as a concept, probably means something different to Einstein than to Bergson. Bergson believes that duration cannot be divided into smaller segments: "our attention may turn away from it and, consequently, from its indivisibility, but when we truly go to cut it, it is as if we suddenly passed a blade through a flame—we divide only the space it occupied."[34] Then, if we add Gurdjieff into the mix, that will bring in the idea of the present. It is only in the present that we are alive. This for Gurdjieff is associated with the idea of *eternity.* This very moment, while you, the reader, are self-conscious of this ongoing now, this is the key to the ulti-mate, your touch with the infinite. Stop for a moment, and consider the *Now*.

Fred Alan Wolf discusses similar ideas in *Star Wave: Mind, Consciousness and Quantum Physics*. Like Penrose, Wolf points out that Einstein and Minkowski have, in a sense, removed "time" by making it the fourth space coordinate. "But the experience of observing can never be mapped. Observation is not an observation. . . . Time is not space."[35] In this sense for Wolf, "Time becomes what we call pure conscious-ness."[36] We can perceive only in the now, and the now happens in time. Wolf is referring to the act of observing, which can exist only in the now, what physicist Amit Goswami calls the collapse of the quantum wave function.

> I have said the brain-mind is a dual quantum system/measur-ing apparatus. As such, it is unique. It is the place where the self-reference of the entire universe happens. The universe is self-aware through us. In us the universe cuts itself into two—into subject and object. Upon observation by the brain-mind, consciousness col-lapses the quantum wave function . . . by acting self-referentially, not dualistically.[37]

For Goswami the experience of the now, that is, consciousness as a process, is nonlocal and thus self-transcendent. According to Charles Musès, "Time is the master control. I will give you an illustration of that. If you take a moment of time, this moment cuts through the entire physical universe as we're talking. It holds all of space in itself. But one point of space doesn't hold all of time. In other words, time is much bigger than space."[38]

Yet the definition of time is filled with subjective connotations. I prefer to define time as *the measure of movement through space,* and the process of consciousness as something that happens in time and is very much linked to Freud and Jung's views on the topic, the structure of the unconscious. Rodney Collins adds the proviso that our time exists "in relationship to a greater world,"[39] in our case, the Sun. One hour is one twenty-fourth revolution of the Earth on its axis, or one twenty-fourth of 1/365th portion of its trip around the Sun. One minute is one-sixtieth of one twenty-fourth revolution of the Earth, and so on. In this sense, time in and of itself is nothing but a ruler of some sort. The essential ingredients in the universe then become *movement of matter in space* (not much different from Hobbes's 1600s idea, which reduced all to motion and matter). The measure of the movement is time. However, movement is also a limiting concept. Time could also be the measure of events occurring in a spatialized progressing sequence. Since to some extent the future contains the possibility for different events to occur, some theorists speculate that time is multidimensional.

The future exists in mind and imagination as potentiality, but it also exists in nature in terms of its teleological component, as seen in the acorn as a potential future oak tree, or in the immune system that is set up to fight future diseases, or in trees losing their leaves in winter so they can withstand the cold to survive until spring.

If we truly want to formulate a more comprehensive model for the so-called spacetime continuum, the question of the location of the mind as an aspect of this construction must be taken into account. Irrespective of human consciousness, aspects of the universe, such as special molecular construction and adaptive mechanisms, already take the future into account: for example, in the inorganic realm, crystal construction as found in snowflakes and geodes promote stability; and in the organic realm, on the macroscopic scale, trees that bend in the wind, leaves that turn toward the sun or convert sunlight, water, and dirt into nutrients, adaptive instincts in animals such as nest building in insects and birds, the burying of acorns by squirrels in preparation for the winter, hunting instincts in predators and flight instincts in prey; and on the molecular level, neurotransmitters that respond to various stimuli, eyes that capture images and send them to the brain, DNA.

AKASA

Whereas physicists have searched for the medium through which light and gravity could travel, occultists throughout modern history have emphasized the relationship of the universal mind to the ether or space, known in Indian philosophy as *akasa*. From the writings of Madame Blavatsky, founder of the Theosophical Society, we learn that "[t]he parent space is eternal, ever present cause of all—the incomprehensible DEITY whose 'invisible Robes' are the mystic root of all matter, and of the universe. . . . It is without dimension."[40]

She says that spirit and matter derive from THAT and it is "neither a limitless void nor conditioned fullness, but both. . . . [Thus] undifferentiated Cosmic Matter . . . is not matter as we know it, but the spiritual essence of matter, and is co-eternal and even one with space in its abstract sense. . . . The Hindus call it *Malaprakriti,* and say that it is the primordial substance, which is the basis of the *Upadhi* or vehicle of every phenomen[on], whether physical, mental or psychic. It is the source from which *Akasa* radiates."[41]

The "akashic records," which are a theoretical mental structure that contains the entire history of the world, can be "read" by psychics such as Rudolf Steiner and Edgar Cayce from past generations and today by John Edward, the early-twenty-first-century TV guru who communicates with the dead.

While the akashic records are an occultist concept, the laws of physics demand that they must exist; and there are numerous ways to prove this beyond such methods involving geology, astronomy, archaeology, and anthropology. Take the case of reverberation. Every sound that has ever been uttered gets fainter and fainter and fainter. But does it ever completely disappear? Physicists would say that the sound diminishes asymptotically. It never totally goes away. Just because we cannot retrieve Napoleon's voice does not mean that it does not exist as some extremely faint vibration embedded in the atmosphere and the walls around where he lived.

Our modern technology has numerous ways to capture the past, such as on newsreel footage, audio- and videotapes, and photo archives. But nature also provides a record through tree rings, carbon dating, fossils, petrifaction, and sediment layers that record major celestial and Earth changes, such as floods, droughts, and en masse extinctions, as when major asteroids have collided with the planet and left their mark as craters.

The proof that the akashic records exist can be established simply by looking at the night sky. One does not see the stars as they are, but rather, as they were. When we look at the stars, we are looking into the deep past. Even our own Sun is seen nine minutes after the fact, because that is how long it takes for the light to travel the ninety-three million miles to arrive at the Earth. At every moment, the Earth radiates itself out to the cosmos, and this radiation contains a precise accounting of the state of affairs of that particular moment. Someone who reads the akashic records is theoretically able to travel back along this time line and see what occurred along the way.

THEORIES OF HARMONICS AND WILL

Louis Acker, a well-known Boston astrologer, in an unpublished paper entitled "Mind: A Holographic Computer," sets out to explain, via a Pythagorean model, how the One God split himself into 2, 3, 4, and so on to create the multitudes. This process is similar to that of vibratory patterns interfering with each other. Just as there are set notes on a musical scale, there are "common nodal points," or "fundamental carrier frequencies," in the creation of the multitudes; a transference of energy from higher dimensions to lower ones can be facilitated by means of principles of resonance and through laws of harmonics. This can be proved on the physical plane with simple tuning forks. All forks with the same dimensions in a room will vibrate if one is rapped. This is the principle of resonance: mutual vibrations. Any tuning fork in the proper geometric proportion to the rapped fork will begin to vibrate as well.[42]

Acker concludes, "Holograms or thought forms existing in the higher frequencies, or superphysical octaves or dimensions [the akashic records and/or universal mind], can reflect themselves down to their sub-harmonic carrier frequencies on the mental astral [human] levels . . . where they can be picked up by a human mind sufficiently organized and calm enough to receive them."[43]

On the other hand, occult writers such as Gurdjieff and Ouspensky tell us that it is the *will*, as an entity unto itself, that provides the transformation from the mental realm to the physical. Other writers such as Carl Jung describe a collective unconscious, or universal mind, from which all individual minds spring and to which all are still connected; every person's thoughts exist at deeper levels (beyond the personal unconscious) in every other person's mind, that is, in the collective psyche. The idea of a collective psyche is a holarchic structure: each thinking mind contains information for the whole and is part of the whole.

Through extension of consciousness into unconscious dimensions and exercise of the will, this domain can be tapped, according to Gurdjieff. Exercise of the will is expression of one's soul, and thus the will becomes an important component of one's identity associated with the biblical name of God: the "I am that I am." The expression of willpower, as Rudolf Steiner profoundly asserts, is directly linked to self-exertion, self-understanding, and expression of one's "I am-ness." Thus, one's link to the Godhead can be achieved through the simple act of conscious doing and the expression/realization of "I am." Since each person is interconnected with the All, mind reading is merely opening up one's own psyche to that which is already present. It is the realization of the inherent self-transcendent qualities within.

LUCID DREAMS

In physics, we speak of energy and its various manifestations,
such as electricity, light, heat, etc. The situation in psychology

is precisely the same. Here, too, we are dealing primarily with energy . . . with measures of intensity, with greater or lesser quantities. It can appear in various guises. . . . As I worked with my fantasies, I became aware that the unconscious undergoes or produces change. Only after I had familiarized myself with alchemy did I realize that the unconscious is a process, and that the psyche is transformed or developed by the relationship of the ego to the contents of the unconscious. In individual cases that transformation can be read from dreams and fantasies. In collective life it has left its deposit principally in the various religious systems and their changing symbols.

CARL JUNG

Besides conscious mental processes, there are other cerebrations that lie outside the normal spacetime continuum; we cannot know them for the simple reason that they are unconscious. As we have seen, according to both Freud and Jung, the unconscious *thinks*. It cannot be over-emphasized how truly profound this insight is. Freudian slips, symbolic behaviors, and dreams all provide evidential support to this hypothesis. Dreams deal with deep inner desires and conflicts. Based on this premise, dreams can be interpreted and decoded to give the dreamer insight. This suggests that the unconscious has its own intelligence separate from the conscious self and that this intelligence oftentimes "talks" to the conscious self, that is, to us, at least the "us" that we think of as our conscious self.

In that sense, we can see the fallacy of the modern cognitive scientists' idea that there is one place for the seat of identity. Clearly, and humbly, one's sense of identity to some extent is an illusion. Gurdjieff is the only theorist I know of who so overtly states this by saying we all think we are one, but really each of us has many "I's," some of which are actually in conflict with each other. As an example, every Friday night I tell myself I will get up early the next morning to get a jump on the day, but when the alarm clock goes off at 6:45 a.m., a different "I" says, "I don't care what that guy said last night, this 'I' is staying in bed for another hour." But Freud is actually saying something rather different here, because these different "I's," for Freud, are not at all conscious. One or more "I's" in the unconscious cogitates. And what is even more amazing, the waking self may have absolutely no inkling at all about the world one or more of these unconscious "I's" inhabits. Anyone who logs his dreams knows exactly what I am saying. There is a whole other world "in there" that the conscious self is oblivious to.

Still, there is the subjective issue that dreams oftentimes feel like journeys to strange and dangerous places. So we are dealing here with the so-called hard problem described in chapter 1. These kinds of dreams feel not like symbolic experiences from the unconscious in the Freudian sense, but rather like excursions to real or perhaps surreal places. To add to the complexity of the point, most dreams never even enter consciousness at all. According to Ken Wilber, the process of dreaming also involves an attempt by the ego to integrate the shadow, not just our "dark side," but also, in the complete Jungian sense, the unknowable, what we cannot know about the process that gives us our consciousness. For Wilber, this is a "path of involution, of return to the source, of remembrance of [primordial] Mind."[44] Some dreams also appear to contain information that arrives telepathically, that is, from someone else's mind.

Primitive societies and the ancients see dreams as true journeys of the soul to what is known as the astral plane. Perhaps some form of reconciliation can be achieved between the two extreme views that (1) dreams are purely manifestations of our own unconscious, and (2) dreams are real journeys to some inner realm. In some way, dreams may indeed be a type of entrance to a different plane of reality where different aspects of who we are intermingle with the souls of other people, both living and dead, and even with other entities.

Where modern conscious humans are, perhaps, several hundred thousand years old, the realm that puts us to sleep at night and creates our dreams is many times older, and in that sense "wiser." Thus, from a pragmatic point of view, it is best to consider dreams as messages from an inner self. In that sense, it is always a wise decision to analyze them according to Freudian methods (see *Inward Journey* or any other major book on dream interpretation to learn various techniques to unravel the psychoanalytic meaning of dreams). Two other key ways to study this realm are to (1) log your dreams, and (2) experiment with lucid dreams, that is, try to take your conscious self into the world of the unconscious and attempt to alter the dream world around you, just as it is altering your "world."

Jung points out, in a way different from Freud, that this inner realm has a mythic reality of its own; it is not just a repository of our fears and repressions. At the same time, dreams also play a role in shaping our conscious activity, from the creation of great art and the divine spark of invention to our struggle with our higher self, our Faustian pact or need to seek a spiritual path to combat the shadow that lurks and lures us down. That is our lot in life.

The psyche can be seen in the Freudian sense as a battle between the animal instincts of the id and the conscience of the superego, with the ego, or self, caught in between. Curiously, it can be seen in a mythic/religious way as the self, torn between the forces of Michael (the angel) on one shoulder and Lucifer (the devil) on the other. Yet ironically, as one of my senior citizen students, Philip Brownell, pointed out in class one night, "Lucifer" comes from the same stem as the word *lucid:* light-giver. In biblical terms, this is the snake, the tempter, the entity that can lead us into darkness or to enlightenment. It is the same force.

Whitrow quotes F. H. Bradley (1902), who analyzed the time sequencing in dreams. This realm, which so easily travels backward and forward in time to reveal past action and future probabilities (and various fantasies), does not really exist in our real outer everyday world and sense of time. "We cannot automatically assume that phenomena exist only if they are in temporal relationship to our world."[45] Yet dreams *may* have a temporal relationship to each other.

Manifestations from *primal* mind, dreams are in some sense untainted by modern civilization. Their modus operandi involves a mental dimension, which, Jung says chuckling, *is* really unconscious, and so, to that extent, *cannot* be understood by the conscious. I know for myself there are many nights that it seems that I have dreamed the entire night, yet when I awake, I can remember almost nothing. Clearly, there is another part of me, which Freud called the unconscious and Rampa, perhaps more wisely, called the *oversoul,* that thinks all on its own.

We barely glimpse the unconscious, this *place* that simply vanishes before our eyes upon awakening. It is a world filled with drama, intrigue, excitement, paradox, violence, horror or delight, complex plot twists, and surreal special effects. Here in the astral planes, there are endless vistas and realms, a picture book of our personal conflicts and fears, the land of the dead, higher and lower beings, past and future places that our inner self seems to return to again and again. We know we have lived in this other place, but we don't know what place that is.

If the dreamer is fortunate enough to have a *lucid dream,* that is, to take a semblance of consciousness or self-awareness into this realm, he or she is often lucky to escape with his or her life, or so it feels. The advanced lucid dreamer learns the laws of this realm and is sometimes able to change an outcome through sheer force of will. But this is rare, and if one does this all the time, then it is no longer a dream, but rather a delusion. The whole point is that the conscious self does *not* control this mechanism. Yet, on the other hand, *the process of lucid dreaming may indeed be the ultimate doorway to a place where the higher entities dwell,* the ultimate realm of inner space. Thus, the goal would be to study the process and techniques of lucid dreaming so that we can learn to bring consciousness into this realm, to change outcomes when possible, and, perhaps more importantly, to interact with the beings we encounter when there.

It may be that we are constructed with a fear barrier of sorts that prevents real access to this inner realm of primal mind. This, for Jung, is an aspect of the shadow, which he sees as profoundly different from the oft cited "dark side." It is unknowable, a part of our psyche that—by its very nature—we *cannot* know. Yet Jung also points out in his autobiography, *Memories, Dreams, Reflections,* that this primal mind influences the march of time and course of human history. People act on dreams in part because glimmers of higher truths pierce through.

FINE MATTER: THE MISSING LINK?

H. C. Dudley, professor of radiation physics at the University of Illinois Medical Center, has suggested that the so-called ether may be a ubiquitous sea of neutrinos, and Arthur Koestler tells us that these neutrinos are not subject to gravitational attraction, nor do they have any mass or electric charge.[46] The absence of "gross" physical properties of such particles as the neutrino, and their quasi ethereal character, encouraged speculations about the possible existence of other particles that would provide the missing link between matter and mind. Physicist Jack Semura (February 22, 2003) e-mails that space might be full of "virtual particles." "Photons," Semura says, "are the carriers of the electromagnetic charge." V. A. Firsoff suggests that "there must exist a modulus of transformation analogous to Einstein's famous equality $E = mc^2$, whereby 'mind stuff' could be equated with other entities of the physical world."[47]

William Lyne, in his book *Occult Ether Physics,* calls particles such as these "omni." "So-called 'empty space' is actually packed almost solid with this very fine matter." Due to its tiny size and very high frequency, "well beyond that of x-rays," omni easily penetrates "solid mass" or flows though mass undetected. Due to its "almost balanced to charge (1:1) ratio, [omni] responds to both positive and negative impulses. . . . At sufficiently high voltages," these quarklike "subdivisions of basic electrical charges . . . are separated by magnetic fields and condensed to form electrons and protons." They are also "interpenetrated omnidirectionally by ultrafine radiation, which is normally in equilibrium, called 'Zero Point Radiation.'"[48] Although next to impossible to find electronically, the omni particle can be detected easily through mechanical action. What we think of as a g-force is actually omni particles "accelerating, decelerating, or changing direction through the atoms and molecules of our body."[49]

CONCLUSION

Physical space may itself be curved, contain antimatter, house a sea of neutrinos, and be related to the invisible realm of the psyche. Nevertheless, physical space is also made up of something else, and that something else has been called, for generations, the ether. With the discovery of holography and a new order to the universe, the conceptualization of empty space must be reevaluated.

Just as the room you are sitting in contains the electromagnetic energy from every radio station and every cell phone conversation in the region, so too does the space within the galaxy contain the light from all of the stars within that region. In continuing the radiowave analogy, it is quite possible that although your radio can pick up broadcasts from stations within a range of only, say, 150 miles, the information from all of the stations around the globe may also be present as some fainter rate of vibration. So, too, may the EM waves from all of the stations from radio history be present as some distant vibration or faint echo. Professor Gora points out that much of the information would be lost in background noise, but he also writes that the first incredible explosion that gave birth to the Milky Way can still be detected!

Gora also tells us that, depending upon where one is within the universe—whether near cloud formations or black holes, near the periphery (like our solar system) or near the galactic core (which may be a vacuum)—the structure of space will vary accordingly. Space is multi-dimensional, housing not only physical dimensions but also the mental conscious and unconscious dimensions. Gora would like to place these unknowns in a fourth spatial dimension called hyperspace; one way or another, by definition, this area will also be part of the holarchy. "There is a concept of hierarchy of spaces," Rudhyar writes, "and space, during a period of cosmic manifestation, actually represents the manner in which all the organized systems of activities operating in any region of the universe are interrelated and interacting. Space is not an empty container into which material substances are poured; it is the interrelatedness of all [bio-cosmic, intra-stellar, and galactic] activities. . . . Operating at different levels of organization or planes of existence, the quality of their interaction and interdependence varies with each level."[50]

One other point to consider is the concept that each person's mind may be a microcosm of the universe. Just as DNA codes for the whole organism throughout every cell within it, the presence of the energy (light) from every star (barring various forms of obstruction) is also present in varying intensities at every sector in space, and thus each spatial sector may code for the whole. A telescope such as the Hubble could be placed at any point within this galaxy or between any galaxies and be able to map the entire vast universe it has access to. Each point is a monad, or microcosm, that reflects the macrocosm.

A map of the structure of the universe is present at every region in space (including the region that our brains occupy). The region of our minds, where this energy is reflected, is thus a form of counter-space, virtual space, or inner realm that is potentially as vast as the outer world. If it is structured holographically, there are also profound implications for such things as intergalactic information exchange and contact with higher beings.

"We can discover a new frame of reference for our new experiences [and technology] . . . in the holistic and hierarchical universe we are coming to know," concludes Dane Rudhyar. "This kind of universe," he goes on, "is being revealed to us because it is the mirror image of what is in us, though still at the stage of potentiality, yet on its way to actualization. Man always discovers outside himself what he is about to become."[51]

When humans learned to place the Earth in its proper relationship to the Sun, many secrets of the universe were revealed. As we progress in our knowledge of our environment, we see that the solar system is part of a larger whole that permeates every level of the biosphere and points us toward a more holistic and self-transcendent view of the galaxy and universe.

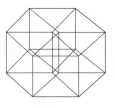

7

COEVOLUTION OF
SCIENCE AND SPIRIT

The Sufi Order of the West, affiliated with Omega Institute for Holistic Studies, presented a symposium titled "Coevolution of Science and Spirit" at the New York Sheraton Hotel the weekend of October 13, 1979. The speakers were Karl Pribram, head of Neuropsychology Laboratories at Stanford University; David Bohm, professor of theoretical physics at Birbeck College, England; Roland Fischer, clinical professor of psychiatry at John Hopkins and Georgetown Universities; and Pir Vilayat Khan, Sufi master and founder of the Sufi Order and Omega Institute. The lectures were structured with panel discussions as well as individual talks.

"A new phase in human thinking has begun," said their promotional mailer. "Starting with theoretical physics and moving into questions of psychological and biological function, science is on the verge of creating a new paradigm which, surprisingly, corresponds to the cosmic view experienced by mystics in meditation."

In thinking back twenty-eight years, I do remember the somewhat informal nature of the event, the easy ability to interact during questions and answers, and the quiet strength of the presence of Pir Valayat Khan, a strikingly handsome charismatic man whose unforgettable gaze rattled one's soul in an amazing, even multidimensional way. I was covering the conference for *MetaScience*. At that time, I had also gained interest in a field that is unfortunately called palmistry. The idea is that the lines, fingerprint patterns, and structure of the hand may indeed reflect brain activity and thus human individualized personality. I had equipment with me to take prints, and as I recall, Andrija Puhrich was there, and since he knew me, he agreed to give me one, and then I got the handprints of Roland Fischer and Karl Pribram.

Not too long after the conference, Pribram went to meet Washoe, the first chimp that could converse with human sign language (who, in the fall of 2007, passed away at the age of forty-two). Pribram, in his attempts to map the human brain had, as a matter of course, done more than his full share of what could only be called gruesome monkey experiments, where he cut out and dyed different parts of their brains and tested them. Thus, he was highly interested in an ape that could literally talk right back to humans. When he reached inside the cage, Washoe, normally a gentle chimp, bit off the top phalange of his middle finger. It is hard to escape the karmic aspects of the event, which did lead to the great neuroscientist abandoning the idea of cutting up monkey brains to figure out how human brains think. Two decades later, at a consciousness conference at the University of Arizona, where we were both speaking (he on the stage, me at a poster session), by pure chance at a luncheon I ended up sitting between Pribram and my doctoral mentor, Stan Krippner. I held my tongue and tried not to look at Pribram's hand. He didn't remember me, I didn't tell him I had a handprint of his with his missing phalange, and we had a pleasant chat.

KARL PRIBRAM:
THE HOLOGRAPHIC PROCESS
OF CONSCIOUSNESS

Karl Pribram led off with the statement that in order to understand the real world, we must understand the nature of consciousness. Reality is something that the brain/mind "constructs . . . but also there is something outside."

In his search for knowledge about the physical world, Pribram was led to the physicists, especially David Bohm and his thoughts on the relationship of holography to the structure of matter, space, and time. As K. R. Pelletier has said in his book *Towards a Science of Consciousness* and Fritjof Capra wrote in *The Tao of Physics,* there is a similarity between the findings of quantum physics and the study of the psyche. At the outset, this may seem a lucky coincidence, but many of the founders of quantum physics actually sought out psychological and Eastern thought for insights into their discoveries. Wolfgang Pauli, for example, discoverer of the exclusion principle and of the orbital and orthorotational properties of electrons, established in his later years a close association with Carl Jung, the well-known psychiatrist and mystic.

Ernst Mach's idea that local properties of space, time, and matter (such as inertia) are determined by the global distribution of matter all over the universe not only stems from Leibniz's idea of how monads interact, but also has its analogous roots in Buddhist philosophy. Like the Buddhists, Mach is proposing that all stars are interrelated; the universe is one. Einstein, a philosopher in his own right, derived much of his thinking on relativity from Mach.

Niels Bohr's formulation of the principle of complementarity, whereby an elementary unit such as a photon acts like a particle *and* like a wave, can be traced to his interest in the Chinese symbol of yin and yang, in which opposing forces are intertwined. Bohr said that rather than choose between the wave theory and the particle theory, we need both of these seemingly irreconcilable descriptions for a full accounting of the properties of photon energy packets.

Werner Heisenberg, discoverer of the uncertainty principle, writes in his book *Physics and Beyond,* "One extreme is the idea of an objective world, pursuing its regular course in space and time, independently of any kind of observing subject; this has been the guiding image from modern science. At the other extreme is the idea of a subject, mystically experiencing the unity of the world and no longer confronted by any objective world; this has been the guiding image of Asian mysticism. Our thinking moves somewhere in the middle, between these two limiting conceptions; we should maintain the tension resulting from these opposites."[1] Here, Heisenberg is attempting to integrate the observer, and thus human consciousness, into the very structure of matter/space-time.

Pribram's ideas on the nature of perception filled in more of the gap. He wrote that there are three major types of explanations about how the brain operates:

- *The neurophysiological dogma:* For every percept there is a particular neuronal arrangement involved with specific neuronal cells. An example is the familiar silhouette against a door window, which is as much Alfred Hitchcock as his face. This theory postulates that perceptions involve an ability to extract features and put them together, but a one-to-one relationship is always implied. A problem with this dogma is that there is a great degree of flexibility in our perceptions.

- *Gestalt theory:* Pribram's mentor, Wolfgang Kohler, supposed that after stimulation of the retina, messages are relayed to the brain, where a direct-current standing electric field is set up. Kohler and

Fig. 7.1. Alfred Hitchcock's self-portrait and signature

Pribram found that these DC fields were isomorphic, that is, they had the same form as the object viewed. However, experiments by Pribram, Lashley, and Sperry showed that even when part of the brain is impaired or destroyed, the ability to recognize an object is not disrupted, even when it is viewed from a different angle from when it was originally perceived. This realization shattered Kohler's theory regarding a one-to-one relationship in perception. Somehow a fragment could trigger a Gestalt response even when the fragment was not isomorphic. Kohler told Pribram, "Now we are really in the soup. We can't understand perception. . . . This has not only destroyed my theory, but everyone else's as well!"

- *Standing waves instead of DC fields:* Karl Lashley theorized that wave-fronts of electric charges and interference patterns, like colliding ripples in a pond, were responsible for reconstructing visual stimuli. Since the brain is roughly 85 percent water, this example relates well to the actual physical structure of our thinking computer.

It should be noted that Pribram purposely simplified Kohler's work for the sake of his talk, because what Kohler and the other Gestalt psychologists learned was that the brain clearly does *not* perceive in isomorphic one-to-one fashion. In fact, the brain is prewired to make suppositions, extrapolate, and form conclusions without the observer being aware that this process is occurring. Note in the illustration above by Alfred Hitchcock how much more is conveyed than just the simple line drawing. Our minds fill in many blanks.

Gestalt psychologists discovered such principles of perception as similarity, proximity, and closure. For instance, when a person looks at the night sky, he or she immediately sees constellations. The reason for this is proximity. Because certain bright stars appear to us as being near each other, it is assumed they are related. In point of fact, they are thousands if not tens of thousands of light-years apart. Other examples are any of the famous figure/ground optical illusions. Is the drawing a young girl looking wistfully into the distance or the profile of an old hag? The point is, people make assumptions all the time based on first impressions and the need for the brain to bring closure, and these assumptions are often completely wrong.

Pribram pointed out that both the brain and a hologram use pulsed frequencies and interference patterns, and in both instances, three-dimensional images are created. In the case of the brain, the outside world has to be reflected/re-created *inside* the perceptual/neurological apparatus of the observer. (i.e., retina and occipital lobe). And then, in a sense, the image is "projected back out" into the world where it belongs. For instance, one sees a tree. The image of said tree has to occur within the confines of the cranium, but the subjective experience, which reflects the reality of the situation, is that said tree appears *outside* the confines of the head. No one truly understands this process, or at least the subjective experience of the process. And to add to the complexity of the example, consider dreams, which unlike outside physical objects are completely inside the head and yet a dream appears to the dreamer as something external to his or her physical being.

As mentioned earlier, even if most of a holographic plate is covered over, the remaining portion can still reproduce the entire image. Similarly, stroke victims who have lost the operative functioning of nearly half their brains can still form images and recall memories. Therefore, it is logical to assume that the brain operates holographically, encoding information in vibratory frequency domains throughout. It is the sensory apparatus (the five senses) that is isomorphic, while the brain is holographic.

Building on David Bohm's concept of a total order being contained or "enfolded"[2] in each region of space and time, Pribram wrote that our spacetime environment is enfolded in frequency domains, spatially within the brain. However, to add to the complexity of this theory, he also discussed the concept of our bimodality of brain functioning in that the left hemisphere, which houses language, predominantly operates sequentially and thus deals with "features," whereas the right hemisphere, which programs art, music, and spatial relationships, is more holistic or holographic in nature. This difference between logic and intuition is also reflected in differences in orientation between the cultures of the East and their interest in spirituality and those of the West and their proclivity toward materialism. He believes we should work to harmonize both domains. Perhaps Niels Bohr's ideas on complementarity could be applied here, in that we need both modes in order to obtain a total picture of our perceptual apparatus and the nature of reality.

DAVID BOHM: THE IMPLICATE ORDER, COSMOS, AND CONSCIOUSNESS

"Matter," said Bohm, "is like a small ripple on a tremendous ocean of energy. The ultimate sources for this world are therefore immeasurable." After asking "What is the hologram of another hologram?" he pointed out that even solid objects are abstractions. The ancient Greeks believed that atoms, by definition, were indivisible. But even the search for subatomic quarks and particles will never, in and of itself, lead to an understanding of the structure of our Einsteinian universe. Relativity presupposes the need for fields rather than particles, but the proposition for the localization of fields is still a mechanistic approach.

When it was discovered that electrons could jump orbits, without going anywhere in between, and Einstein, Podolsky, and Rosen discovered their principle of nonlocality, whereby information seems to traverse space instantaneously (that is, faster than the speed of light), it was deduced that even things at great distances from one another were connected in a special way. Bohm suggested that whereas relativity relies on causality, continuity, and locality, quantum physics deals with the exact opposite: acausality (Heisenberg's uncertainty principle), discontinuity (orbital jumps), and nonlocality. Bohm set himself the task of unifying these complementary theories, seeing that both quantum physics and relativity propose no real divisions in the universe; each is just constructed from a different perspective. According to him, instead of dividing the world into its parts to explain it, we need a new notion of order, which he calls the "implicate" (or enfolded) order.

First off, Bohm said, we must consider that the world is always in a state of flux, an unbroken flow of energy transformations. He suggests that objects are unfolded out of this holomovement. The separateness of objects is a secondary (explicate) order deriving from the deeper primary (implicate) order. The explicate order is a particular order that interpenetrates and intermingles with the implicate order. The fundamental enfolded order has relative stability and recurrence features. Since everything derives from the unbroken flow, or holomovement, nonlocal connections or quantum jumps are possible.

If we are talking about perception, it occurs to this author that the explicate order could refer to the outside world and the implicate order to the mind/brain mental realm, and they intermingle during the simple process of perception and more complicated processes of cogitation and volitional activity.

Even though the discovery that Einstein made with Podolsky and Rosen suggested instantaneous information transmission, Einstein couldn't "stomach" that conclusion. He felt instead that this discovery proved that quantum physics had some type of internal flaw, because nonlocality was impossible. Other physicists, however, have ignored Einstein's caveat and accept the actuality of nonlocality.

Bohm used the following example to demonstrate how our observations on the structure of the universe limit its dimensions in proportion to our comprehension. He drew a picture of a tank containing two fish, then asked us to imagine two television cameras, each one recording the image of one fish and projecting it onto a separate screen, with the result that a viewer would see two separate fish in two different places, when in actuality they are both in the same tank. The universe has a hidden interconnectedness, but it is obscured by, among other things, the consensus reality of our times; as a result, we do not see the unifying factor in all things. We cannot, for instance, see the relationship between chance and destiny, yet many people would argue that there is one. According to Bohm, there is a "higher dimensional reality," which allows "relatively autonomous subtotalities" (such as the current laws of physics), but there are greater or "deeper" laws. "Current cosmology sees the beginning of the universe as an explosion of space [rather than matter]; therefore our universe is a small ripple of something else. When considering consciousness, we see that matter is enfolded in mind." "Descartes," Bohm went on, "said that matter is extended substance, and thought is not. This is a theory that has taken man away from true understanding, for God has created them both." Consciousness and matter are implicate orders of a more comprehensive or more fundamental order.

Consciousness, primarily in the implicate order, involves a display in the explicate order. "Our higher dimensional reality" Bohm suggests, "includes both consciousness and matter enfolding each other. The time sequence is also enfolded, and is a display of another implicate order. Processes in time unfold from the implicate. This can be seen graphically in evolution. The next stage is not completely related to the previous stage, but is also related to a higher order. Creativity allows new content at each moment, but we must also remember that thought can not come from absolute truth." This last statement is rather ponderous, in that it implies that humans can never comprehend reality.

It may be beneficial to interject here Gurdjieff's ideas of evolution and involution. The first involves the striving for the organism to become something more (expression of implicate order), and the second is what Gurdjieff calls "help from above." An animal could not evolve if the higher aspect was not already present as a potentiality. I would see this as a higher or more fundamental level of implicate order (e.g., Bergson's élan vital): birds fly because flying is possible. The explicate order would be linked to Darwin's theory whereby the environment forces or makes demands on the implicate order—adaptation and survival of the fittest: polar bears are white so that they will blend in with the snow, moths look like fallen leaves, and so on.

DAVID BOHM AND KARL PRIBRAM: THE NEED FOR A PARADIGM SHIFT

Bohm and Pribram discussed the need for a new construct of reality to better explain the mysteries of consciousness research. Bohm brought out the importance of the structure of language. We not only communicate with language, but are also limited by it. Therefore we must learn to better define the term *consciousness*—by use of our language (see opening chapter). As the Russian neurologist A. R. Luria has said, the very manipulation of mental concepts increases neurophysiological capability. According to Luria, the development of human language also allows the psyche to utilize symbols in place of physical objects, thus creating new associative patterns. Greater neuronal complexity ensues. A person can imagine, draw a blueprint, plan a future, and this in turn alters and creates more advanced states of consciousness.

Pribram discussed the need to differentiate between consciousness and self-consciousness; he also pointed out that in scientific studies, repeatability (of an experiment) is only a small part, whereas understanding is more important. He referred to Lashley's remark that a situation becomes interesting when it does not follow the norm. "If the Israeli psychic Uri Geller can psychically bend a key only once," Pribram said, "it still is important and should be understood."

Bohm suggested that the holomovement flows and never repeats, like two hot days in August that may be similar but are not identical. "The end result of the Einstein, Podolsky, Rosen discussion on non-locality is that we are all different, yet all of one mind." Bohm's term sheds new light on old ideas, offering another way of defining the universal psyche, the oversoul, the collective unconscious.

ROLAND FISCHER: ALTERED STATES OF CONSCIOUSNESS

Fischer told us that he had experimented in 1945 with mescaline while in Switzerland. His travels through the world of psychotropia led him to realize that there is biological knowledge that is quite different from physical reality. "Is man separate from the universe because he observes it?" he asked, making the point that—just as Einstein made use of an observer to explain his theory of relativity—cosmology must address itself to the fact of consciousness as one of the properties of the universe.

Fischer referred to the search for ultimate truth recorded in ancient Hermetic manuscripts, which contain thirty-four definitions for God, including: "God is a sphere of which the center is everywhere and the circumference is nowhere." In this symbol of consciousness, even the center remains infinite. Hermeticism teaches that the inner dimensions can be known only through direct experience. Thus the search for higher states can be accomplished only by each of us for ourselves using our perceptions, which Freud said arise from awareness. However, psychotherapy tends to ignore the metaphysics of inner knowing by concentrating on a person's "physical" experiences.

Fischer discussed the principle of identity in perception. "The laws of nature are the laws or our own nature. If I were not Sun-like, I could not perceive the Sun. We must return to the Pythagorean view," he suggests, "and see God in the center of higher states of consciousness. Observer and observed can not be taken as separate entities. . . . Whatever you look at, you are always looking at your own brain. Consciousness expands as internalization of experiences proceeds, and language does its best to code what it perceives."

PIR VILAYAT INAYAT KHAN: MEDITATION AND HIGHER STATES OF CONSCIOUSNESS

Pir Vilayat Khan concluded the seminar. He first spoke about his father, Hazrat Inayat Khan, who had come to the West in 1910 to bring forth the Sufi philosophy. "Tracing its roots back almost 10,000 years, the Sufi teachings not only contain many secrets regarding highest states of consciousness, Sufiism also lies at the base of both Buddhist and Judeo-Christian teachings."

Pir Vilayat presented a unique charismatic combination of spiritual knowledge and objective scientific thinking. Representing "the man of experience," he said, "to explain God is to dethrone God [because] one tends to reduce Him to a concept. When one is really experiencing this reality, one finds no words to express it."

During meditation one experiences cosmic and transcendental dimensions. "By extending one's consciousness into the vastness, one loses identity and merges with the totality. [Further,] each element of the totality contains another totality within it. . . . The totality of spirit transcends the logic of the mind. . . . Meditation," he went on, "is a state of resonance. You feel in total contact with the universe even though you don't know what the universe is!"

Integrating Mach's principle of universal interconnectivity, Jung's concept of the collective unconscious, and Kammerer's thoughts on seriality (synchronicity, or meaningful coincidence), Pir Vilayat suggested that our mental world transcends the physical. However, even on the physical plane, "our bodies are made up with the substance of the planet. [Even] DNA, formed from the beginning of time, contains

Fig. 7.2. Pir Vilayat Khan

the whole history of the Universe . . . but we must watch not to over-simplify."

Having been "beyond the beyond," Pir Vilayat discussed the elusiveness of the true self. "Who is the 'me' that makes the lump of flesh move?" he asks. [If we] watch our consciousness, are we not our consciousness? Consciousness is just like a flame. It is dependent upon fuel and content. By realizing that we are pure intelligence, we enter into divine consciousness, merging, as in deep sleep, with everything of a transcendent nature. Once one experiences that state, one can never be the same."

Going even further in integrating Western thoughts with Sufi teachings, Pir Vilayat observed that "matter" is rapidly experiencing an extraordinary upsurge of intelligence. As Koestler has written in his description of the hierarchy, "Higher wholes have structures, which could not be predicted from lower levels. The random state of entropy is really a state between two orders. But for this greater order to take place, randomness also has to take place.[3]" Khan told us, "As a Sufi, I believe that 'I am the instrument through which He experiences Himself.'" Khan told us to reflect upon the nature of our own selves, and to accent the purpose of being rather than the cause. "My father's teaching is to make God a reality. Don't look for God up there," he said pointing above. "He is here."

"Samadhi is a state of unity, that is, the consciousness of unity behind the diversity. . . . In this state, when you are in a forest, you experience what it is like to be a tree. You become the tree. There is a famous Sufi saying: 'I look at him with his eyes and he looks at me with mine.' This ancient thought relates directly to Jesus's discussion of the most important teaching from the Old Testament: 'Love thy neighbor as thyself.' And Moses' attempts to explain the deeper meaning of monotheism. We are all a part of the same consciousness," Khan concluded, "and therefore we must integrate our purpose with 'divine purpose,' in order to create and build a wonderful world."

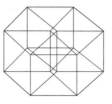

8

MONAD OF MIND

All physical circumstances take place in the material world. According to traditional scientific theory, the speed of light is the limiting velocity for the occurrence of these material events. However, the mind clearly transcends time. It can reflect on the past or future and exist in the virtual realm of imagination or in the many dimensions of the unconscious. In that sense, it may somehow travel "faster" than the speed of light, or at least exist in a dimension where the speed of light is not a limiting or meaningful factor.

Properties of mind permeate the structure of matter, including the elementary particles, because electrons and protons display some primary form of sensitivity and self-activation. Somehow, the electron is "aware" it "likes" protons but is repelled by other electrons. This built-in sensitivity is associated with a fundamental aspect of consciousness. Something is directing the electron. This something must be both intrinsic and extrinsic. It is part of undifferentiated mind, and present also in our brain.

Mind is intimately linked to the concept of light: "seeing the light" and its equation with consciousness are common motifs in our culture. In this regard it is interesting to note that all matter is held together because photons are shared among atoms. This is a law of physics. Light energy—which is somehow related to consciousness—is the glue that holds matter together. Perhaps photons hold the key to the secret of the concept of mind. If they do, then insights may be gleaned if we begin to reflect on how these photons are spatially arranged, their gestalt, so to speak.

The structure of our brain and neurotransmitters both must play a role in this arrangement. It is known, for instance, that there is an input pathway, via the optic nerve, for light or photons to enter the brain to stimulate the pineal gland. This gland—which philosophers like Aristotle thought was the seat of the soul—produces serotonin and melatonin, two neurotransmitters closely related to the processes of becoming conscious and dreaming. The entire brain is involved in most (if not all) conscious activity. (Pulling one's hand away from a hot stove happens subcortically, so it can be argued that this is a form of consciousness or awareness that the body has irrespective of the brain. The event, however, will end up being registered in the brain.)

If we merely look at a chocolate-chip cookie, the occipital lobe at the rear of the brain receives the visual input, the language center in the temporal lobe labels it "chocolate-chip cookie," intentions are considered in the frontal lobes ("should I eat it, or will it ruin my dinner?"), and the parietal lobe (left angular gyrus) coordinates the lobes and thereby augments the sizing of it up for possible confiscation.

Our brain works holistically in a "kinetic melody of concertedly working cortical zones."[1] This holistic property of the brain has been further explained by Nobel Prize–winner Roger Sperry and his colleague Michael Gazzaniga, who have discovered double endgramming, where information set up for processing in one hemisphere is actually placed in both hemispheres, and considerable crossover between the two hemispheres occurs. Although the left brain processes language, the right brain can secretly understand verbal commands. And although the right brain is the one that draws the pictures, the left brain can certainly comprehend what a picture is. Thus, the left brain can think in nonverbal terms and the right brain can understand verbal commands.

For the sake of simplicity, we will ignore these secret abilities of one hemisphere to do what the other one is programmed for, because this added complexity makes it more difficult to understand the nature of the brain's superstructure.

THE STRUCTURE OF THE BRAIN

In the early 1950s, Sperry intrigued the scientific community with his "split-brain" studies performed with individuals with severe forms of epilepsy. He knew that the electrical "storms" created by the disease in one hemisphere would cross over to the opposite hemisphere to cause mirror-image electrical damage at the same spot on the opposite hemisphere. Sperry hypothesized that if he cut the corpus callosum, the link between the two hemispheres of the brain, the epilepsy could no longer occur. Severely epileptic patients thus had the connecting links between the two hemispheres surgically severed. To the amazement of the scientific community, this procedure worked.

Sperry went on to test each hemisphere of these epileptic split-brain patients separately. He found that each hemisphere thinks inde-

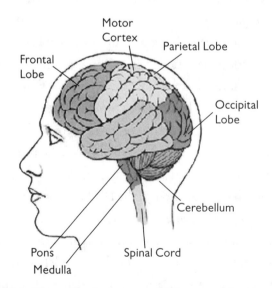

Fig. 8.1. The cerebral cortex, cerebellum, and brain stem

pendently of the other. One famous experiment performed by Sperry (1961) drives home this point.

When a split-brain patient was asked to draw this figure,

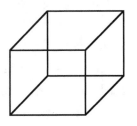

Fig. 8.2. The target

the right brain drew this:

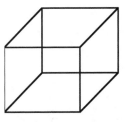

Fig. 8.3. What the right brain drew (simulated)

Where the right brain drew a credible replication of the square, the left brain drew something like this:

8.4. What the left brain drew (simulated)

By testing each "brain" individually, Sperry discovered that the left brain, which houses language, thinks temporally and sequentially, and the right brain, which is nonverbal (or perhaps preverbal) and is involved in face recognition and spatial concepts, thinks holistically and intuitively.

Since the left brain thinks sequentially, that is, in piecemeal fashion, when it saw a line it drew a line, it saw another line and drew another line, and so on. The right brain, on the other hand, comprehends holistically, so it saw and drew the entire box. Below is a drawing that depicts the well-known differences between the left and right hemispheres of the brain. The left houses language and technical skills, thinks logically, and sees the parts, whereas the right programs pictures and music, thinks intuitively, and sees holistically.

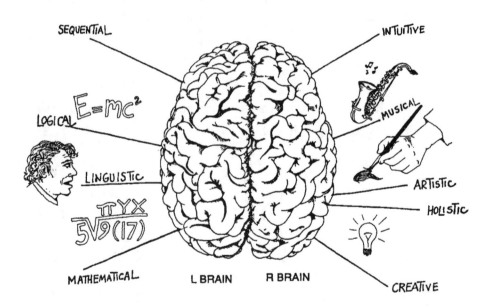

Fig. 8.5. Bilateral mind drawing by Lynn Sevigny

Simplifying this model, we come up with the following:

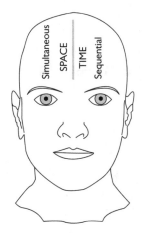

Fig. 8.6. Space/time mind

These extreme simplifications help to demonstrate some fundamental differences between the two hemispheres. To continue with the discussion, we see that our frontal lobes make use of logic and thus also think sequentially, whereas the occipital lobe, situated in the back of our head, which houses the visual cortex, deals with spatial relations. It thus sees holistically, that is, "simultaneously." A. R. Luria, the

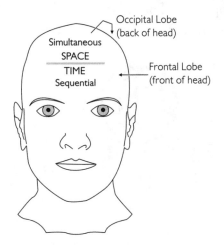

Fig. 8.7. Time/space mind

acclaimed Russian neurophysiologist, has stated that the parietal lobe, also in the rear of the brain, coordinates the temporal and occipital lobes and thus deals with what he called "simultaneous synthesis."

If we simplistically view the superstructure of the frontal lobes as compared to the parietal and occipital lobes, we arrive at the following diagram:

```
SIMULTANEOUS
I               E
M               Q    Right Brain      Left Brain
U               U    wholes           parts
L               E    pictures         words
T               N    art              science
A               T    intuition        logic
N                    music            language
E               I    simultaneous     sequential
O               A
U
SEQUENTIAL
```

By combining the two complementary models of left/right and front/back "brains," we come up with the following "monad of mind."

The top of the box refers to the back of the brain, or occipital/parietal region, and the bottom of the box refers to the frontal lobes.

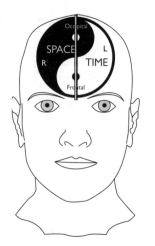

Fig. 8.8. Yin/yang mind

Seen as facing us, the left side of the box refers to the right hemisphere and the right side refers to the left hemisphere. Much like Niels Bohr's principle of complementarity, whereby a photon is both a wave and a particle, the basic unit of consciousness might have sequential and simultaneous aspects.

Fig. 8.9 is a visual reminder of the intersection of time and space and shows that our minds perceive both temporally and spatially.

THE BRAIN AND THE STRUCTURE OF THE COSMOS

We are transducers and organizers of conscious energy radiating from all the stars, and our brain arranges these myriad photons in a highly structured electrophysical spacetime network. This neurological grid is also possibly a holographic reflection of the cosmos. It is truly composed of all the forces of the universe, containing light from all its stars. Each of us is a miniature universe and each reflects the whole.

In general, we can pair the left brain to "consciousness" and the right brain to the "unconscious." Perhaps our brains reflect the cosmic interplay between day and night, Sun and Moon, just as they are split between logic and intuition, conscious and unconscious. Consciousness would be linked to daylight and thus the Sun, and the unconscious would be linked to the night and the Moon. Taking this a step further, the brain is 85 percent water and therefore highly susceptible to subtle changes in electromagnetic fields. Just as the tides are affected by the Sun and the Moon, it is possible that the day and night circadian cycle of the brain hemispheres, monitored by the pineal gland, may also be affected by these two heavenly bodies. During the day, the Sun rules, and our conscious left brain, which thinks in words, is dominant. At night, when the Moon is out, a reversal takes place, and we sleep and dream. The left brain, which was conscious, becomes unconscious, and the right, which was relatively unconscious, becomes "conscious."

One of the functions of sleep is to place the mind in a state whereby the day's events can be cataloged. To accomplish this, the conscious mind must be asleep. The theory is that this occurs during the REM cycle. We know that dreams often reflect the day's events. It takes a lot of energy to be awake and to process information. That is a key reason why students often fall asleep in the library, why people are exhausted after visiting a new city for the first time, and why infants spend so much time in REM. In each case a significant amount of newly acquired information has to be electronically stored. Information that has been accumulated in the mRNA in the cell body of the neurons (messenger RNA, which holds the day's events in short-term memory) must be transferred to the corresponding different parts of the brain.[2] Motor memory will be stored in the motor cortex, visual memory will be placed in the occipital lobe, neurolinguistic memory goes into the temporal lobe, and so on. This transferring of information from mRNA to long-term memory located at the ends of the dendrites most likely takes place when we dream. A person who learns a lot in life will actually have a more complex brain and more dendrite connections than a person who learns very little. Purged of this information in the morning, or after a powerful nap, the mRNA can be ready again to take on new data.

THE SENSE OF I

Although some theoreticians want to place the sense of identity in a particular locale—Francis Crick, for instance, suggests the claustrum, an off-shoot of the amygdala that is involved with coordinating cerebral processes—Nobel Prize–winner Eric Kandel states that it is by no means clear how the brain creates the "unity of consciousness." He suggests that it is more likely that the sense of self is linked more globally to frontal lobe/thalamic interactions.[3] Located in the center of the midbrain, the thalamus is, in a sense, the Grand Central Station of the brain. All input first goes to the thalamus and then is rerouted to whatever lobe the information is supposed to go to. If the thalamus is damaged, the person loses his or her sense of identity.

The role of the thalamus in combining the two separate ways of thinking of the two hemispheres of the brain is not totally understood. The thalamus, however, is directly related to the numinous subjective sense of self. Combined with the total brain in an evolutionary sense, it is a culminating structure that allows, in some way, for an aspect of the universe to observe and reflect upon itself. Gurdjieff would go so far as to say that one of the key roles for the human is to be a conscious mechanism for the evolution of the cosmos.

The process of becoming conscious, or being conscious, is further complicated by the fact that the brain operates with an alternating current. As you read this page and are awake, your mind is not at all aware that the field that is generating cerebral activity is alternating its electronic charge at somewhere between fourteen to sixty-plus cycles per second. This is to say, in a mere second, your brain has changed its charge from positive to negative to positive to negative at least fourteen times as rapidly. Yet the subjective feeling of a single consciousness persists. Put another way, one does not detect a flicker. This persistence of a stable sense of awareness during such a complex electronic process may have application to this new etheric view of the universe we are constructing, whereby our very substrate, the ultimate energy that compromises the All, is also oscillating at a very rapid rate, but when we look out at the physical world, those ubiquitous polar reversals remain undetectable.

Properties of mind permeate everything, but when the mind is highly structured, as in human beings, it becomes an individualized thinking entity. It remains part of the whole but is also a volitional monad that is both self-contained and interpenetrates all other minds.

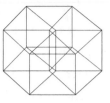

9

SYNCHRONICITY

In this [subatomic] world of moving things, you can imagine it is something like a great game of chess played by the gods. [You may think that you have discovered all the rules], but every once [in a while] something like castling is going on.

RICHARD FEYNMAN,
FIVE EASY PIECES, 1961

In 1909, the well-known zoologist Paul Kammerer completed a book on coincidence entitled *Das Gesetz der Serie (The Laws of Seriality).* It contained exactly one hundred coincidences, which Kammerer had collected over a twenty-year period. Kammerer studied each occurrence in much the same way he would classify any species, such as "the lizards of the Adriatic Islands."[1] He would then delineate and analyze each coincidence according to such criteria as the number of similar occurrences and the amount of shared attributes. By studying them in this way, Kammerer hoped to uncover the underlying structure and mechanics of synchronistic phenomena. Arthur Koestler discusses one of the more spectacular coincidences in his biography of Kammerer, *The Case of the Midwife Toad:*

During the holiday season of 1906, Baroness Trautenburg, a spinster born in 1846, was injured by a falling tree, and at a different place Baroness Rigershofen, a spinster also born in 1846 was injured by a falling tree.[2]

In this case the number of similar occurrences is two and the number of shared attributes is four: baroness, spinster, age, falling tree. The precision of the synchronistic occurrences becomes more apparent when they are broken down this way. Koestler informs us that many scientists were intrigued by *The Laws of Seriality*. In reviewing the book, Einstein commented that the work was "by no means absurd."[3]

The systemization that Kammerer used to categorize coincidences comprised six types of series:

1. Homologous and analogous
2. Pure and hybrid
3. Inverted
4. Alternating
5. Cyclic
6. Phasic

For instance, once, while driving on the highway, I saw four Volkswagen Beetles, by chance, strung across the four lanes. If, at that moment, I was listening to a Beatles song, that would correspond to an analogous series as there are four musicians in the group: John, Paul, Ringo, and George. If years later to the day, I saw four yellow cars strung out on the highway in a similar format as the Volkswagens, this would be, according to Kammerer, a mystic hybrid series.

Kammerer also discusses the "morphology" of "non-causal" occurrences:

1. The order is the number of similar events.
2. The power is the number of parallel occurrences.

3. The parameters are the number of shared attributes.

The central idea is that side by side with causality of classical physics, there exists a second basic principle in the universe, which tends toward unity, a force of attraction comparable to universal gravity. But while gravity acts on all mass without discrimination, this other universal force acts selectively to bring like and like together both in space and in time; it correlates by affinity, regardless [of] whether the likeness is one of substance, form or function, or refers to symbols. The modus operandi of this force, the way it penetrates the trivia of everyday life, Kammerer confesses to be unable to explain because it operates ex hypothesis the known laws of causality. But he points to analogies on various levels where the same tendency toward unity, symmetry, and coherence manifests itself in conventionally causal ways.[4]

The psychoanalyst Carl Jung, heavily influenced by Kammerer and by acausal models that were being developed by such quantum physicists as Werner Heisenberg and Wolfgang Pauli, continued with this line of study. He coined the term *synchronicity,* which means simply "meaningful coincidence," and carried the analysis of synchronis-

Fig. 9.1. Carl Jung, circa 1959

tic phenomena into speculations on the structure of the psyche. Like Kammerer, Jung described a number of astonishing examples:

In April 1949, I made a note in the morning of an inscription containing a figure that was half man and half fish. There was fish for lunch. Someone mentioned the custom of making an April Fish of someone. In the afternoon a former patient of mine showed me some impressive pictures of fish. In the evening I was shown a piece of embroidery with sea monsters and fishes on it. The next morning I saw a former patient who was visiting for the first time in ten years. She had dreamed of a large fish the night before.[5]

A few months later, when Jung was writing up this series, he walked down to the lake and saw a big dead fish upon a sea wall. Jung sought to explain events such as these as evidence of a connecting principle in nature that transcends our common notion of space and time, cause and effect.

Causality is the way we explain the link between the two successive events. Synchronicity designates the parallelism of time and meaning between psychic and psychophysical events, which scientific knowledge so far has been unable to reduce to a common principle. The term explains nothing, it simply formulates the occurrence of meaningful coincidences, which, in themselves, are chance happenings, but are so improbable that we must assume them to be based upon some kind of principle or on some property of the empirical world. No reciprocal causal connection can be shown to obtain between parallel events, which is just what gives them their chance character. The only recognizable and demonstrable link between them is common meaning or equivalence.[6]

A famous coincidence occurred in March of 1979 when Columbia Pictures released the movie *The China Syndrome* within two weeks of the nuclear disaster at Three Mile Island. Not only did the fictional story and actual crisis appear virtually at the same moment in time, but also there were other uncanny similarities.[7] In both instances:

1. The accident took place after a pump on the generating system shut down.
2. Gauges malfunctioned.
3. There was a distinct possibility of a meltdown (*The China Syndrome*).
4. Human error played a significant role.
5. Officials tried to minimize the danger that was quite real.
6. In the movie there was a line that said that a meltdown in California could "render an area the size of Pennsylvania permanently uninhabitable"; in actuality, the real crisis occurred in the state of Pennsylvania.

The relationship between the movie and the actual event provides an example of what Kammerer termed an "analogous" series. Both Kammerer and Jung felt that amazing coincidences were not simply chance events and that the laws of nature operate at deeper dimensions than the mere surface levels of everyday events. Kammerer thought that the laws of coincidence concerned principles similar to that of gravity, whereby similar events would be attracted to each other. Jung suggested that coincidences were manifestations of the human collective consciousness. At deeper layers of an individual's mind, he or she taps into a "universal psyche" shared by all. Within it are psychic structures that not only transcend individuals, but also transcend time and space. Jung called these *archetypes,* formative patterns that move across time, having the ability to influence seemingly unrelated events and able to manifest in more than one mind at the same time. He believed they tend to attract events around themselves.[8]

Synchronistic phenomena . . . prove that a content perceived by an observer can, at the same time, be represented by an outside event, without any causal connection. From this it follows either that the psyche cannot be localized in space [and time] or that [spacetime] is relative to the psyche.[9]

The psyche is not only the container of the personality, but is also the connecting principle between mind and matter. Just as a pebble thrown into a still pond spreads its waves out in a circle, archetypes in the realm of the mind may initiate and send out connecting energies to different parts of the physical environment. This model can be represented by the Jungian mandala with an initiating archetype at the center, its influence radiating out like the spokes of a wheel to each physical event at the periphery. In this manner, an unconscious force from the central layers of the collective psyche may cause similar events to manifest in the physical world. Although separated on the physical level, the events are connected at deeper levels.

Applying this theory to the case of *The China Syndrome* movie and the similar real-life occurrence of the accident at Three Mile Island, for example, neither event caused the other, but both were "caused" by something else, an autonomous initiating archetype. Thus, Jung's "acausal" model may really point to one of the more fundamental (beyond spacetime) causes. Certainly, if the collective psyche can be established as a scientific reality, then the scope of our modern-day philosophies would be significantly affected, as an overarching mental realm would be shown to intercede with the physical.

Critics of synchronistic phenomena believe that astonishing coincidences are simply "short-run illusions of a longer hidden series."[10] Their argument essentially is that given the hundreds of billions of everyday human events, a certain natural percentage of them should be coincidental. As the well-known magician the Amazing Randi told me one day, "[I]f coincidences did not occur, then that would be something!"[11] A closer look, however, certainly offers much to consider. Nevertheless, the novelist Kurt Vonnegut suggests that synchronistic phenomena may in fact occur, but he warns us to be wary of "granfalloons," or plain ordinary accidental coincidences that are mistaken as meaningful![12]

SOME PERSONAL EXAMPLES OF SYNCHRONICITY

Throughout the mid-1970s, I made part of my living as a jewelry designer and salesman. I was living in Rhode Island, and my parents still lived in New York, where I was raised. This was about 170 miles away by car. My mother, a Brooklyn College mathematics major, worked as an accountant at a jewelry manufacturing company on Long Island. Through this company I obtained the components necessary to create my designs. Modestly successful, I decided one Friday to expand my territory, and so drove eighty miles north to Marblehead, Massachusetts, and then on to the seaport tourist town of Rockport, which is fifteen miles farther away, just north of Gloucester. Rockport has, perhaps, one hundred shops, and in one of them I sold my entire line. Elated, I drove home.

By coincidence, this was the same day that my parents had decided to take a trip to Canada, where my father was born, which was about 350 miles from where I was. Unknown to me, on a whim, my parents decided to go to Rockport instead of Canada. Meandering through the myriad shops that same Friday, my mother purchased an item as a present for me. When she went to pay for it, she noticed all of the jewelry in the case by the cash register. Upon inquiry, she learned that the owner had just obtained the line from a young man from Rhode Island. "That's my son!" she exclaimed. We missed each other by about an hour.

The following year, in 1979, I had a vivid dream about Wilt Chamberlain. As I awoke, I reflected upon the dream, when suddenly the clock radio went off. One of the first things the radio announcer said was that Chamberlain had been elected to the Basketball Hall of Fame. He had been retired for many years, and I could think of no particular reason that I dreamt of him. I have encountered this type of dreaming interlink with the radio or TV on several other occasions as well.

The extreme precision that can be shown by this type of phenomenon is depicted in the following example. One Saturday in 1975, my girlfriend asked if we could drive up to a factory outlet in Brockton, Massachusetts, so that she could shop for sweaters. I agreed, and took along a book on parapsychology to read while she rummaged through the racks. While sitting in the store and reading, I came across a passage that contained the word *pharaonic*. When I discussed the word with my girlfriend, we decided (correctly) that it must be linked to the word *pharaoh*. Since there was a museum nearby and we had some time, we drove over and entered. Down in the basement was an Egyptian exhibit with a beautiful mummy case with a pharaoh inside. That, in and of itself, seemed to me to be just a coincidence. However, that evening, while we were watching the Dick Cavett talk show, he used the word *pharaonic* in a sentence! This astonished me, as before that day I had never even known that such an adjective existed.

While reworking this chapter, I was also trying to locate the cell phone number of a friend who was visiting from out of town. The day before, I had been in Providence on an unrelated matter to meet with a lawyer who worked for a large firm. He said that he thought that he may have hired me a number of years before, but neither of us could remember whether this was so, although I did think I had worked for his firm as a consultant. I pulled out a six-year-old date book to try and track down the cell phone number of the friend, because I knew that I had first met him about that long ago. On the same page as his cell phone number was the name of this same lawyer written in another-color ink!

Another interesting synchronicity occurred December 10, 2007. I have two mailing addresses. At one address I received a letter from my mother, which contained a full-page review of a new play on Broadway about Philo Farnsworth, the inventor of the TV tube. From the other mailbox I retrieved a letter from a gentleman recalling that it has been ten years since I had spoken on Tesla at the New York Public Library. There was a new lecture on this topic there, and would I be interested in attending? The letter was signed J. Bart Farnsworth! This letter was from a man I had never heard of, although I may have met him ten years ago. I, of course, knew who Philo Farnsworth was.

In the instance concerning Wilt Chamberlain, the dream preceded the airing of the story on the radio. This, however, would not be precognition, because the most likely explanation is that my mind either picked up the thoughts of other radio listeners who had heard the radiocast one or two hours before I awoke (which would be telepathy) or my mind had tuned in to the actual radio frequency itself (clairvoyance). Telepathy requires the interlinking of two living minds, and, by definition, clairvoyance involves obtaining information psychically from matter, or, in this case, FM frequencies. The synchronistic event involving my mother purchasing a gift from a store where I had just sold jewelry could also be a form of telepathy. She would have had to have subliminally picked up my going to that particular store. The other event, involving the word *pharaonic,* is more difficult to explain.

I want to reiterate that telepathy may be a much more common phenomenon than we think. This morning, November 17, 2007, I awoke with a dream about a homeless lady whom I discovered staying in our basement. She had broken in, and was sleeping on a couch that was given to my wife, Lois, right after she left college. Lois had it in various apartments through the 1970s, and we had placed the couch in the basement of this house when we moved here in 1980. Twenty years later, we finally threw it out. At breakfast this morning, my wife asked me if I had read "Ask Amy" yesterday. I said no. It was a story about four college girls who rented a house off campus. One of the four had agreed to rent sight unseen and ended up in the one room in the basement, which didn't even have a window. She asked Amy what she should do, as it was clearly an inferior room. Amy suggested she find another place to live. Telepathy with my wife is a reasonable explanation for the source of my dream.

A synchronistic experience of a different sort occurred on another occasion in 1975, while I was preparing the graphology portion of a lecture with an astrologer for a local library. We planned, respectively, to analyze the handwritings and horoscopes of Adolf Hitler and John Kennedy. I chose these two individuals because I had copies of their handwriting. The astrologer was able to find Kennedy's horoscope, but was unable to locate Hitler's. Therefore she decided to draw up the horoscope herself.

The lecture was rather unsuccessful because only one person and a newspaper reporter showed up. With so much time to spare, we walked over to a stack of about ten graphology and astrology books that the librarian had kindly collected for us. The astrologer picked up one of the astrology books. "Oh, Noel Tyl, he's an interesting author," the astrologer said, referring to a person I had never heard of. With those words she randomly opened the book to Hitler's horoscope!

Now, this synchronistic experience does not seem to be explained by any known form of ESP. Even if she clairvoyantly or subconsciously knew that Hitler's horoscope was in the book, the odds were still hundreds to one that she would have opened to the correct page. In actual point of fact, counting all of the books collected, there were about five thousand pages to choose from.

Another synchronistic experience I had is also not so easy to explain by any potentially known mechanism. On April 19, 1976, I had a dream of a multicolored two-dollar bill. At that time, two-dollar bills, although rarely used, had again been placed into circulation. The bill in the dream was red, blue, green, and yellow. Later that day, I went to the grocery store and purchased eggs, milk, and bread and the bill came to exactly two dollars. I reached into my pocket and to my total astonishment pulled out a two-dollar bill and paid the grocer with it! I had no idea there was a two-dollar bill in my pocket.

The night before, my girlfriend had returned three dollars to me, which she had borrowed. Without looking, I took what I thought to be three one-dollar bills and placed them in my pocket. Unknown to me, she had, of course, for fun, paid me with a one-dollar bill and a two-dollar bill. Needless to say, the pulling out of this bill at that time was an amazing experience.

A couple of amazing synchronicities have occurred during these last few weeks as I have been preparing a final draft of this manuscript. The first one took about two years to materialize. I am a handwriting expert with over thirty years' experience. In all that time, I had never reported a lawyer to the Bar Association for refusal to honor a debt. After many months of phone calls, letters, complaints, and so on, I was finally forced to report a lawyer to the law board, and the case was resolved in my favor. The lawyer's name was uncommon, the same as that of a Hollywood star from the 1940s, so let's say it was "Grable." About six months later another lawyer with the same name from a different state hired me in another forgery case. I was immediately aware of the connection in names, but proceeded with the work and wrote a long, detailed report based on the material. Amazingly, this lawyer also did not pay my full fee. Perhaps in part because of the potential synchronicity, I handled the situation in an extremely patient manner, calling occasionally and sending monthly bills. However, after nine months of nonpayment, I was left with no alternative and so contacted the Bar Association in his state. This second "Grable" was so mortified that he sent a bonus as a way of apology. So, the only two times in more than thirty years that I have been forced to go to a law board to report a lawyer involved two unrelated lawyers with the same unusual name.

The other synchronicity occurred the weekend of July 5, 2007. I was considering adding to this text a quote from Jane Roberts. I went into my archives and retrieved *Seth Speaks,* a book I do not think I have opened in possibly twenty years. I looked through the book and saw all the underlining I had done back in the 1970s, but could not find a quote that fit my needs. I placed the book back where it had been buried. The following night, a girlfriend of an associate of mine who had passed away came over to spend the weekend with us. One of the first things she did when she came through the door was say that she was reading *Seth Speaks*. She reached into her pocketbook and retrieved the very same purple paperback that I had put away less than twenty-four hours earlier.

THE KICKER

I have made a discovery in synchronistic phenomena that I call "the kicker." It is an extra coincidence that, like the icing on the cake, is one more attraction that makes an ordinary synchronicity truly spectacular. The kicker to the *Seth Speaks* synchronicity is that during the week (October 22–28, 2007), as I was reworking this chapter, I met a book agent online who told me in a phone conversation without any prodding from me that she is the agent that represents *Seth Speaks*. The kicker to the astrology episode was the fact that less than a week after the incident and completely to my surprise, at a New Age conference, I met the author of the book the astrologer had pulled out at the library, Noel Tyl, a man I had never heard of before! I was so taken by the experience that I recounted the entire story to him. The kicker indicates both extreme precision and a mystical element of synchronistic phenomena.

SYNCHRONICITIES INVOLVING MOTHER NATURE

In 1980, a natural disaster coincidence was reported by ABC Sports. During an ice skating trial, a story was told by one of the skaters about a tornado that had descended upon the ice rink in Chicago. The girl who related the tale said that she was able to get off the ice in time, but her friend did not and was killed. The next day one of the dead girl's skates was found eight miles away in the backyard of her grieving parents! If this story is accurate, it is a most astounding occurrence of a correspondence between the Earth's natural disasters and individual human events.

In January of 1996, there were five serious train accidents across the nation involving derailments and, in each case, death. In Rhode Island, that same week, a serious oil spill occurred off the coast at Moonstone Beach. For Rhode Islanders, this event was particularly tragic because the accident occurred at the epicenter of what many consider to be the most pristine natural reserve in the state. Thousands of dead lobsters washed up on the beach, and many seabirds as far away as Block Island, ten miles from the site, were also destroyed.

The spill, which was composed of more than 800,000 gallons of a quickly dispersed heating oil, although tragic, was small as compared to a series of non-dispersible crude oil spills that had occurred in New England waters twenty years earlier, during the last two weeks of 1976. On December 15, the oil tanker *Argo Merchant* ran aground off the coast of Nantucket. Two days later the 1970s remake of *King Kong* opened its New England premiere "featuring a story of an oil company's discovery of the ape while searching for oil and the catastrophic result of their commercialization." Six days later the *Argo Merchant* broke up, dumping a staggering 7.6 million gallons of heavy crude oil into the sea. The following day a tanker spilled 25,000 gallons of crude in Narragansett Bay in Rhode Island. Within a week, another 2,000 gallons were spilled from yet another tanker into the Thames River in Connecticut, and on the last day of the year, December 31, the *Grand Zenith* went down south of Nova Scotia. This ship was carrying 8.2 million gallons of crude.

In this New England region—where no major spills can occur for years, even decades—we had five major oil spills within a two-week period. The kicker, in this case, occurred on December 18, when the *Providence Journal* published a photograph of the *Argo Merchant* stranded on the shoals, with its serial number, 382, featured prominently on the hull. Three Rhode Islanders played that number in the lottery and it came in, paying off $2,500 apiece.[13]

In a similar vein, on September 11, 2002, a year to the day after the terrorist tragedy at the Twin Towers in New York City, the number 911 won in the New York State Lottery. Ironically, the winnings per person were very small because so many people played that number that day. In both instances regarding the lottery, a psychokinetic component is possible at the level of physical causation, as so many minds were cued into those numbers at those two separate times. Thus, when the numbered Ping-Pong balls pop up, perhaps, sometimes a mental energy of sorts ends up influencing the outcome.

Just as this book was going to press, the country was startled by the sudden death of the notable newscaster, Tim Russert, host of *Meet the Press*. Only 58, Russert was highly respected. His memorial service, held on June 18, 2008, was attended by numerous VIP's, including two presidents and presidential candidates Barack Obama and John McCain, who sat together in church. After the ceremony, the attendees went outside, and saw above them a spectacular double rainbow. The image was so breathtaking that even hard-nosed reporters who covered the service on national TV, were astounded. The mystical synchronicity of the event simply could not be overlooked.

TELEPATHY AND ESP

Telepathy may be a much more common form of communication than we think. Personally, I have found that some of my dreams can be explained by a telepathic model. For instance, I have had the experience of dreaming about someone I had not been in contact with for a long while, only to find a letter from that person in the mailbox the following day.

ESP may be responsible for some of our actions in a way that is similar to how other actions are initiated from the unconscious. For example, in April of 1976, I had a day off and journeyed down to New York City to do some work at the Society of Psychic Research. Traveling by train, I decided to take along the book *My Story*, Uri Geller's autobiography, as reading material. Engrossed in the book, I missed my stop on the subway. Rather than crossing the tracks to switch trains to go back to the stop I missed, I decided to emerge at Columbus Circle and walk the fourteen blocks along Central Park West to the society's headquarters.

About six blocks down, strolling toward me, were John Lennon and his wife, Yoko Ono. We exchanged greetings, and then I said, "John, since I have your photo on me, would you mind signing an autograph?" Lennon was surprised by this statement, so I clarified. "It's in Geller's book," I said, pointing to the photo of the two of them conversing at a table. This coincidence allowed me to meet John Lennon and Yoko Ono, obtain their autographs, and analyze their handwriting. It is possible that some form of telepathy would explain the encounter, as I may have unconsciously known that Lennon was in the area.[14]

BIRTHDAY SYNCHRONICITIES

One of the male students in my spring 1976 parapsychology class at Providence College night school had the same birthday as his grandmother and aunt (August 8). A female student's grandfather, father, and brother, all only sons, were born on February 7, 1898, 1924, and 1947, respectively. I have also noticed coincidences regarding my own birthday, February 17:

- I was born on the same day as the woman who introduced my parents to each other, with whom they remained friends throughout my upbringing.
- I grew up in a small town with another boy named Marc who, like me, spelled his name with a "c," even though it was a very uncommon spelling. Further, we were both born on the same day of the same year, February 17, 1948.
- When I moved away from Rhode Island to Chicago in 1972, my

close friend RK befriended another male student, DF, who had the same birthday as mine. In many ways he replaced me as RK's closest friend.

- While teaching a parapsychology class in 1978 at the University of Rhode Island extension school in Providence, I was lecturing about birthday synchronicities and randomly pointed to a girl and asked her for her birth date. She replied February 17, which, of course, is my birth date. The following day, I repeated the lecture at another college, and then randomly pointed to a girl in that class and asked her for her birth date. She said February 17! This second event astounded me and was clearly the kicker.

- My writing partner for eighteen years on the Tesla screenplay project, whom I met by chance, Tim Eaton, has the same birthday as I do (see chapter 2).[15]

In my high school yearbook, I wrote that my goal in life was to be a pro basketball player. If you'd ever seen me play the game, you would realize this was a joke! However, I do share a birthday with one of the greatest basketball players who ever lived, Michael Jordan, who was three years old when I graduated high school.

In the case of close relatives born on the same day, it is possible that cosmic forces such as the rhythmic path of the Earth may play an important role in determining optimal days for conception and birth. We know, for instance, that other animals have optimal times for mating. It may be possible that certain humans are also sensitive to seasonal influences. Some of the other birth date synchronicities mentioned above are not so easily explained.

SYNCHRONICITY WITH TWINS

The history of parapsychology is replete with coincidences between identical twins reared apart. At the University of Minnesota, the psychologist Thomas Bouchard has performed research in this area. Being a cautious scientist, Bouchard has not attempted to relate the similarities to any "parapsychological" causes. He seeks a more materialistic basis for the phenomenon. One of his coworkers, David Lykker, suggests that the genes may be responsible "in determining all aspects of behavior."[16]

Two half-Jewish twins reared apart, one as a Nazi in Germany and the other in the Caribbean as a Jew, share the following peculiarities: "The twins like spicy foods and sweet liquors, are absent-minded, fall asleep in front of the TV, think it is funny to sneeze in a crowd, flush the toilet before using it, store rubber bands on their wrists, read magazines from back to front and dip buttered toast in their coffee."[17] The kicker in this case might be the flushing of the toilet before using it.

The case of the "Jim twins," reared apart and unknown to each other until they were adults, is one of the most famous instances of twin synchronicity (see *Inward Journey* for a detailed account). This case has perhaps five or six kickers. Both boys were named Jim, both married a Linda, both divorced her to marry a Betty. Both had dogs named Toy. Both built a seat that half wrapped around a tree in their respective backyards, named their sons James Allen and James Alan, and died on the same day!

Although this is a remarkable case, it is not unique. Astounding similarities with twins reared apart are somewhat common occurrences. Since identical twins have developed from the same egg, the structures of their brains are highly similar, essentially mirror opposites. Thus it seems possible that identical twins may share more of a "common mind." The telepathic link could be so close that if one twin falls in love with a Linda, another may experience the feelings as well and therefore seek out a Linda to satisfy the feeling. Keep in mind that these two Jims not only have identical brains, but they also "grew up" together for nine months in the womb, and for four weeks after that, even though they were then separated. Later, as they developed into children and eventually adults, it is quite possible, if not probable, that each twin was aware in some primordial, unconscious way of the existence of the other. Some form of resonance or mutual vibrations most likely created a sympathetic channel between them.

The suggestion by Lykker that genetic determinants may be responsible for the twin coincidences[18] appears to be a forced hypothesis, although genetic factors, no doubt, play a role. If part of our psyches includes the atomic structure of DNA, then it is reasonable to consider theories that search for a physical substratum for the detection, transmission, and source of information transference.

PSYCHODYNAMIC THEORY AND SYNCHRONICITY

A possible explanation about how the twin synchronicities could occur involves a synthesis of the models of Freud and Jung. Each person has a conscious and a subconscious. Freud separated the subconscious into a preconscious, which contains latent (just below the surface) memories, and a personal unconscious, which contains repressed and forgotten memories. Beneath this layer, according to Jung, is the collective unconscious, which involves a dimension of mind common to all humans.

FREUD AND JUNG'S MODEL

Conscious—awareness of internal and external states

Preconscious—latent memories, the censor (keeps painful memories buried in unconscious), automatisms (habits)

Unconscious—forgotten and repressed memories, the real self

Collective Unconscious—the foundation of the mind, instincts, racial memories common to all humans

Fig. 9.2. This table starts with the simple Freudian model of conscious, preconscious, and unconscious and places the collective unconscious beneath, thereby synthesizing the models of Freud and Jung. The deeper portion of each individual mind has transcendent qualities.

In the case of identical twins, part of their personal unconscious may telepathically link them with each other. This is depicted in the table below.

TWIN A	TWIN B
Conscious	Conscious
Preconscious	Preconscious
Unconscious	Unconscious
Shared Personal Unconscious	
Collective Unconscious	

Fig. 9.3. All individuals would share the collective psyche. But identical twins may also (telepathically) share a percentage of their personal unconscious. From the point of view of physics, their minds would be in a resonant frequency for the simple reason that their brains are anatomically identical.

This idea is closely associated with Bruce Taub-Bynum's concept of the family unconscious, which would lie between the personal unconscious and the collective unconscious. Realizing that family life helps shape the dream life, and further that personal ego boundaries are weakened by sleep, Taub-Bynum suggests that the collective psyche of a family is to some extent an open system whereby information can flow telepathically from one member to another.[19]

Paranormal information apparently arises through the same psychic apparatus as normal memories when appearing in consciousness. They all come up from the unconscious through the same pathway, that is, through the preconscious to awareness in the conscious. This can make it difficult to distinguish a paranormal thought from a normal one. Similarly, paranormal information arising in a dream may suffer distortion by the censor, such as appearing in symbolic form, similar to that which occurs in normal dreams. This makes it difficult to realize that a dream may, in fact, have been telepathic.

However, in contradistinction to the mechanisms of symbolic representation, some forms of synchronicity, such as the meaningful coincidence, tend to utilize great precision of duplication, such as shown by the twin cases, birthday synchronicities, the two-dollar-bill incident, and when odd words come up, as in the case of the *pharaonic* series. In these instances, there seems to be no symbolic distortion whatever.

The Microcosm

Paul Kammerer saw seriality or synchronistic phenomena as comparable to gravity, where there would be a symmetrical tendency to bring together like and like. The ultimate unity of the cosmos is alluded to, and it can be explained by a variety of models, both psychospiritual and material.

The mandala concept discussed above, for instance, can be expanded to place the "I am that I am" in the center. Radiating out, stemming from innermost space like expanding rings, would be first the initiating archetypes, then the collective psyche, the family unconscious, the personal unconscious of each individual, the preconscious, and finally, on the outward ring interacting with physical matter, an awareness and self-awareness function called by Freud "the conscious."

This Freudian/Jungian mandala model links all people at the source. At the periphery we are individuals; at the center we are one. David Bohm suggests that this ultimate mental realm would have an implicate and also an explicate order, unfolding and enfolding. Information can emanate out from the interior or information can be absorbed back into the center from the outside. There are no real divisions in the universe, according to Bohm. Rather, each part is just constructed from a different perspective. Synchronicity is very possibly a mental counterpart to the physical process of holography. Somehow the same event is distributed (sometimes in derivative form) to many regions. The whole is found in all the parts.

POSSIBLE MECHANISMS FOR
SYNCHRONISTIC PHENOMENA

One theory for transmission involves extremely low-frequency (ELF) electromagnetic (EM) waves.[20] This theory attempts to explain synchronistic and telepathic phenomena as having a purely physical substratum. In this case, the mechanism and carrier wave would be various electrical oscillations that pervade the Earth, which are also synchronous to human brain waves. "The ELF-EM field-waves and impulses occupy the frequency band between 3 HZ and 3 KHz while (Ultra Low Frequency) EM phenomena (less than 3 Hz) occupy adjacent bands."[21]

These data suggest that geomagnetic pulsations directly relating to the brain-wave frequencies may act as carrier waves for various kinds of ESP phenomena. Since these waves are always surrounding the Earth, it would also be theoretically possible for telepathic information to travel at the speed of light. This velocity would approximate three and a half revolutions of the Earth per second, which for all intents and purposes would be instantaneous. Since alpha EEG waves (at 8 Hz) have been correlated to relaxed states, this suggests that ESP occurs more often during alpha production, a frequency, Persinger notes, that is identical to "Shuman resonances," which create standing waves surrounding the Earth's ionosphere.[22] "Psi information signals," according to this theory, "are carried on extremely low electromagnetic frequencies to which temporal lobe structures are sensitive."[23]

If one considers the structure of the brain itself, there are a few other possible mechanisms of transfer. One to consider would be the electronic frequency of the brain associated with its sizable water content. Individual molecules of H_2O are held to each other by an H-bond, which is a weak attraction between the positive side of one molecule and the negative side of another. It is the rapid H-bond shifts, occurring at about one million times per second, that account for the fluid property of water. It is very possible that this H-bond shift frequency, which would also be in harmonic resonance with water molecules in the air or ocean, could act as a carrier frequency for direct brain perception (DBP), that is, the transfer of information from one brain to another. Certain neurotransmitters could be in synch with this frequency.[24]

Other possibilities could be the very structure of the carbon ring, which may promote some type of nonlocal resonating channel, or quantum tunneling, as proposed by Walker (1977) and Goswami (1993). We know that telepathy often occurs during emotionally charged times and during dreaming, that is, when certain neurotransmitters, such as adrenaline, dopamine, and serotonin, are being produced. The molecular structure of all of these neurotransmitters contains a carbon ring and various combinations of other atoms of carbon, hydrogen, nitrogen, and oxygen. The carbon ring is a hexagram-shaped, field-generated structure composed of electrons, protons, and neutrons. Viewed as a multidimensional frequency configuration, the ring comprises six carbon atoms symmetrically arranged, with six nuclei, each surrounded by six electrons circling in their various orbits.

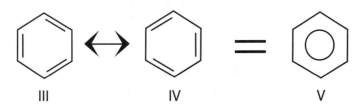

Fig. 9.4. Two carbon rings connect to each other by resonance to form benzene. This resonance model was hypothesized by Linus Pauling in 1930. The resonant energy helps stabilize the molecule.

An atom's nucleus, which contains protons and neutrons, was initially thought to be very large. Now physicists know that the nucleus of an atom is exceedingly small. For instance, if the orbit of the innermost electron were the equivalent to the circumference of a sizable room, the nucleus, according to Richard Feynman's illustration, would be a bare speck of dust in the center of the room. The nucleus is thus, in a sense, "deeper inside space" than the orbiting electrons (which themselves are spinning on their own axes). At the present time, it is still accepted that electrons (and photons) are "elementary particles," that is, they do not appear to be composed of smaller components. The protons, however, are composed of three quarks, which, like the nucleus, can be seen as embedded even more deeply in space. In any event, these elementary and subatomic particles may potentially be seen as doorways to other realms where our thoughts and the collective psyche dwell. It is here, in this subatomic realm, linked to inner space or hyperspace, where the fabric of normal spacetime may be pierced.

If we compare the human brain/mind to the computer model, we can see the brain, our bodies, and the physical world as associated with the "hardware," that is, the actual computer. Normal communication would be analogous to sending information along transmission lines and via wireless communication. The actual information, equivalent to the mind, would be associated with the "software," and cyber or virtual space, a realm that can be accessed by every single computer linked to the Web. This model can be seen as analogous to Jung's idea of a collective psyche. One way or another, synchronistic phenomena involve the idea of mutual vibrations. Resonant effects and the principle of harmonics may also be utilized to access dimensions that transcend the so-called physical plane of reality.

DISCUSSION AND CONCLUSION

Through brain-wave resonant effects, identical or similar information could arise in the consciousness of two individuals who have in common neurophysiological characteristics. This model would support the finding that coincidences tend to occur between identical twins with more frequency than in the general population. Family members, or individuals who share an emotional bond or interest, would also be more likely to have synchronistic phenomena occur between them.

Telepathy, as a form of synchronicity, may occur with greater frequency on unconscious levels than on conscious levels, as through dreams and supposed chance encounters. Transfer of information may also take place without the person or people involved even knowing that synchronicity has occurred. For instance, if I had *not* listened to the radio the morning that I dreamed about Wilt Chamberlain, I would not have known that my psyche had possibly picked up information in an extrasensory fashion. Or I might have *had* the dream and not remembered it on a conscious level upon wakening! If this is the case, telepathic transfer may be occurring all the time, but because humans are aware of such a small percentage of their unconscious processes and their dreams, they remain unaware that this transfer may even be taking place.

In the case of *The China Syndrome*/Three Mile Island synchronicity, or birthday synchronicities among generations, the initiating factor stems from deeper strata that transcend our normal concepts of space and time. Generations come and go, but the psyche of the human species lives on. This, in essence, is the collective unconscious where Jung's archetypes dwell. Existing as an all-pervasive hyperspatial realm in its own right, the collective psyche too can initiate synchronistic phenomena in the same way that one twin can initiate it in the mind of another. If an initiating archetype becomes manifest, it can apparently trigger similar events in different individuals, locations, and situations. Resonance and harmonics would play key roles, as the two separate instances that are seen as coincidental are linked for the very reason that they are similar, and recognized as such.

Synchronistic phenomena may be mental manifestations of a process similar to holography, whereby each separate event is somehow linked to some central unifying source. Neurologically, the carbon ring may play a key role as liaison between physical and mental domains. The study of synchronicity may lead humans to uncover heretofore-unknown layers of the psyche. Its study may also shed light on the relationship of the structure of the human mind to everyday events and the physical world.

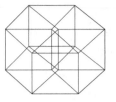

10

PRECOGNITION AND THE STRUCTURE OF TIME

Throughout history there have been a number of civilizations that have produced individuals who have accurately predicted the future. The Bible, for instance, is replete with prophecies and predictions. Note this quote from Isaiah relating to the life of Jesus (Isaiah 41:25):

> I have roused up one from the north, and he is come; From the rising of the Sun one that calleth upon My name; And he shall come upon rulers as upon mortar. And as the potter treadeth clay, a harbinger of Zion will I give: Behold, behold them; And to Jerusalem a messenger of Good tidings; And I look but there is no man; Even among them . . . there is no councilor; That when I ask of them can give an answer.

In fact Isaiah is filled with predictions of the upcoming Messiah. In the forty-fourth verse, that a name appears quite clearly resembles Jesus:

> Yet now hear, oh Jacob my servant, and Israel, whom I have chosen, thus saith the Lord that made thee and formed thee from the womb

who will help thee; Fear not, oh Jacob My Servant, And through Jeshrum whom I have chosen.

The ancient Greeks had one of the most prophetic societies. They erected oracles, which were actual temples where gifted psychics and priests met to describe the future. The best-known oracle was at Delphi, where a large stadium and place of worship was erected and decorated with beautiful gardens and "innumerable statues of bronze and gold." The holy men looked to the stars for their knowledge of the future and combined their astrological data with the predictions of a "priestess," a psychic medium who went into trance by breathing in psychotropic gases and drinking sacred water. While in this altered state, she would see the future. Often her information was symbolic, but sometimes actual events were foreseen.

Cicero said of the Delphi oracle that it "never could have been so overwhelmed with significant presents from all kings and nations, had not all ages proven the truth of its oracles."[1] The Oracle at Delphi was only one of many such ancient prophetic institutions that existed for the uncovering of messages from the future, along with soothsayers like Spurinna, whose famous prophecy "Beware the Ides of March" prefigured the death of Caesar.

In the fourteenth century, Pierre d'Ailly (1350–1420), chancellor of the University of Paris and astrological counsel to the king of France, predicted the French Revolution of 1789: "Should the world still remain in existence at that time [1789] which God alone knows, then astonishing upheavals and transformations will occur, which will effect on laws and political structure."[2] A century later the astrologer Nostradamus (1503–1566) followed in d'Ailly's footsteps, but soon took over as the world's greatest seer (see chapter 12 for more on his work).

Several concepts and practices relate to precognition, as shown in figure 10.1, page 208.

CONCEPTS OF PRECOGNITION

Foretelling	Déjà vu
Foreshadowing	Entelechy
Forecasting	Teleology
Foreknowledge	God's will
Foreordained	Tomorrow, yesterday, and today
Future	Free will
Past	Chance
Present	Randomness
Probable futures	Indeterminacy
Possibilities	Unpredictable
Preordained	Chaotic
Premonition	Acausal
Prophecy	
Prophet	Omen
Prophetic	Oracle
Prediction	Seer
Predestined	Geomancy
Predetermination	I Ching
Precognition	Tarot
Astrology	
Deduction	Palmistry
Fate	Clairvoyance
Destiny	Divination
Spacetime	Writing on the wall
Time	Feeling in my bones
Time symmetry	Asymmetrical time
Causality	Backward causation
Periodicity	Backward time
Cycles	Reincarnation
Synchronicity	Eternal recurrence

Fig. 10.1. All of the terms above bear a relationship to the concept of precognition. They have been grouped in general clusters of similarity of nuance.

In the modern era precognition, the child of prophecy, became a reputable area of investigation. J. B. Rhine, father of modern parapsychology, established essentially single-handedly that the ability to accurately see the future could sometimes occur. Rhine used simple laboratory experiments with a deck of cards (although he was not the first scientist to do this). Rhine simply had subjects intuitively predict the order of a pack of cards *before* they were shuffled. The predictions of a number of individuals invariably scored significantly above chance, whether the cards were shuffled and laid out five minutes, one day, or a month later. Using statistical analysis, Rhine provided scientific evidence for the existence of this phenomenon.

Another important feature of his work is that it can be easily replicated (including by this author on many occasions). I encourage you, dear reader, to try this yourself. Get a deck of cards. Line up a few willing subjects and perform these experiments. This certainly is not the only way, but it is a good way to study this phenomenon in a controlled scientific environment. Further, it is the difference between belief and knowledge, skepticism (or faith) and direct experience.

PRECOGNITION AS A FORM
OF TELEPATHY

In July of 1979 I had a dream about being at the Navy boatyard along Narragansett Bay at Quonset Point, Rhode Island. I was standing at the wharf throwing tennis balls up at a huge aircraft carrier that was docked there. The next day I went sailing with a friend. While waiting at the dock for him to bring in the rowboat, I saw two young boys begin to throw rocks at the yachts. The similarity to the dream was immediately apparent. I had never seen other boys throw rocks at yachts. Given that this is a paranormal experience, would it be precognitive?

With the supposition that both myself and these boys knew ahead of time that we were going to be at the boatyard at that particular time, it is quite possible that I telepathically perceived the plot to throw rocks as it arose in the conscious or unconscious of (one of) the boys. I may have received this information telepathically due to the possibility that the boatyard created a psychic link between myself and these boys and the forthcoming event. This theory of a simple resonance model for thought transference has been called "factor G" by Persinger (1979). He speculates that if thought is vibration, resonance effects must take place; much like the operation of gravity, a local region in space (such as the boatyard) can cause disturbances affecting the psyches of different people in similar ways. The boatyard may have acted like a carrier frequency. Although this was a precognitive event for me, it was probably a predetermined event for the boys (that is, they were planning on throwing rocks the day before they actually did).

The criteria for "pure precognition" clearly must include components that have (for all intents and purposes) aspects related to the event that could *not* have been known ahead of time. Since telepathy with the boys was a possible explanation for the rock-throwing incident, it would not be a case of "pure precognition."

A famous case often cited as precognitive is Lincoln's dream of his own assassination a few days before the actual occurrence.[3] However, telepathy could also account for the dream in that Lincoln could have "tuned in to" the mind of one of the men who were plotting to kill him. A large number of the laboratory studies and spontaneous occurrences of precognition could be explained by psychokinesis, telepathy, clairvoyance, subliminal perception, or conscious or unconscious perceptions and subsequent subliminal deductions and predictions.

THE SEARCH FOR PURE PRECOGNITION

Researchers Robert Morris and Judy Kesner have attempted to create a pure case of precognition. They had twenty subjects relax and imagine a scene that would be matched to a record album to be randomly chosen at a later date. The following week a number was selected from a stock page (shares-sold column) and used to select the album. Independent judges scored two direct correlations with odds for chance around 100/1. Whether this study actually involved precognition cannot be completely ascertained. At least it was a creative attempt to design a test for precognition, including the essential criterion of components that could not have been known ahead of time.

In my spring 1975 parapsychology class at Providence College, a lady stated that she was psychic and had witnessed, in a clairvoyant vision, a well-known airplane collision "in her living room" the day before it happened. As such an event seems to be completely unpredictable, this would fall in the category of "pure precognition."

A tarot card reader in a subsequent class allegedly predicted the striking of a tree by lightning a week before the occurrence. Such an incident contains an element of prediction that could not be explained by any form of logical deduction.

In an attempt to establish whether or not pure precognition actually exists naturally, two cases in the literature stand out. Each involves the writing of a book that predates and predicts an event with uncanny similarity. Neither author had attempted to predict anything. Therefore, the following instances are evidence for the concept of precognition as a human potential, but not proof that humans can consciously predict otherwise unknowable futures.

Case #1: The *Titanic*. Fourteen years *before* the sinking of the *Titanic*, novelist Morgan Robertson (1898) wrote a novel entitled *Futility*, concerning the sinking of a huge steamship called the *Titan*. His fictional liner was considered unsinkable but foundered after crashing into an iceberg. There were a number of distinct similarities between his story and the later event, which might indicate the paranormal nature of this episode in history.[4] And, in fact, before the real ship left port, a few individuals decided not to board it. Dean reports that their reasons were due to premonitions concerning the safety of the boat.[5]

Stevenson (1970) reports one case of a New York woman who was awakened by a nightmare concerning her mother being a passenger on

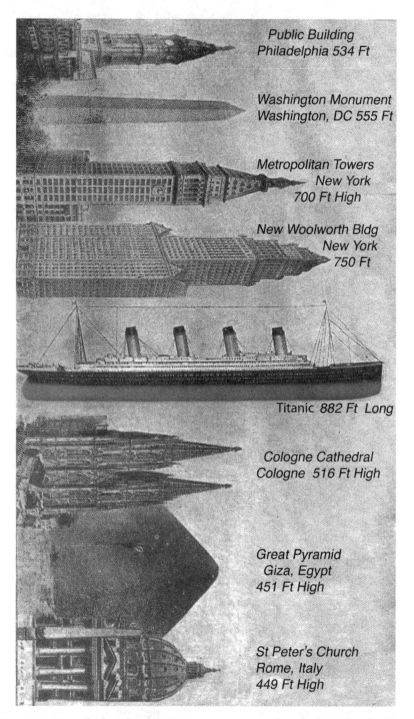

Public Building
Philadelphia 534 Ft

Washington Monument
Washington, DC 555 Ft

Metropolitan Towers
New York
700 Ft High

New Woolworth Bldg
New York
750 Ft

Titanic 882 Ft Long

Cologne Cathedral
Cologne 516 Ft High

Great Pyramid
Giza, Egypt
451 Ft High

St Peter's Church
Rome, Italy
449 Ft High

Fig. 10.2. Note the length of the Titanic *as compared with the tallest buildings of the day. The sinking of this ship was a tragedy of mammoth proportions. The psychic event concerning the 1898 story about the fictitious ship the* Titan *adds a prophetic mystical variable comparable to, perhaps, an ancient biblical tale.*

a lifeboat in the mid-Atlantic. She reported to her husband that she could actually feel the cold and salt spray on her face. Her husband reassured her that her mother was safe. However, it was later learned that the woman had booked a surprise passage on the *Titanic*. The dream occurred two days before the actual sinking of the ship.

To explain this *Titan/Titanic* episode as a chance coincidence seems to me to be a weak argument. It could be argued that the man who built the boat read the book. This could account for a number of the coincidences, such as the name of the ship and the problem with the lifeboats. It is hard to argue, however, that the reading of the book could have caused the ship to hit an iceberg. Nor does it seem likely that a form of psychokinesis was responsible for the accident. The table below demonstrates the synchronistic nature of this precognitive episode. As Kammerer would suggest, there was a recurrence of a similar event.

	TITAN	TITANIC
Number of people aboard	3,000	2,207
Number of lifeboats	24	20
Speed of impact	25 knots	23 knots
Displacement of tonnage	75,000	66,000
Length of liner	800 feet	822 feet
Number of propellers	3	3
Month of accident	April	April
Hits an iceberg	yes	yes
Sinks	yes	yes
Many people die	yes	yes

Fig. 10.3. The similarities to a fictional story written in 1898 and the actual sinking of the Titanic *in 1912 are extraordinary. This table delineates most of the comparisons.*

Interestingly enough, if the *Titanic* had hit the iceberg head on, it probably would not have sunk. The ship was designed with five independent sections. Even if two had been punctured, the ship would have stayed afloat. The iceberg, however, sliced through three of the sections under the hull. It was because the ship nearly missed the iceberg that it sank! This type of grazing impact was an astounding bit of "bad luck." Therefore, the fact that many people died in both the story and the actual event was also a rather unpredictable coincidence. The ocean liner was equipped with a wireless transmitter and under different circumstances, a reserve ship might have saved the day.

Case #2: Patty Hearst. This case also involves the coincidence of a story in a book with the actual event. In 1972 James Rush Jr. (pseudonym Harrison James) wrote a book entitled *Black Abductor.* A young college student named Patricia, daughter of a wealthy and prominent man, is kidnapped near her campus where she is with her boyfriend, who is severely beaten. Initially, the boyfriend is a suspect in the case. The kidnappers, led by an angry young black man, are members of a terrorist revolutionary group. At first the girl is an unwilling captive, but later adopts their ideology and joins the group. They send Polaroid pictures to her father along with a message in what is termed America's

Fig. 10.4. Patty Hearst involved in a bank heist while still a kidnap victim

first political kidnapping. The fictional abductors predict they will ultimately be surrounded by the police, tear-gassed, and killed.[6]

Black Abductor appeared in print before Patty Hearst was kidnapped. Like the *Titanic* synchronicity, the events were so similar that *Black Abductor* would have been prophetic even if the girl's name had been other than Patty. As in many synchronistic events, there was an extreme precision of recurrences.

Here, also, we see the relationship of synchronicity to precognition. If we look for alternative explanations, it could be argued that the black revolutionary leader of the Symbionese Liberation Army read the book (or that the book was written based upon research into militant American revolutionary organizations). This would explain some of the similarities. However, this case has an unusual ("unpredictable") twist, which is that Patty Hearst ended up joining the group. We certainly would not have expected this from someone so deeply entrenched in a wealthy and traditional way of life.

The kicker in the *Black Abductor* instance is the use of the same name, Patty. With the *Titanic* episode, we have several kickers, such as the name of the ship and the name of the captain being the same as well!

These two cases, although occurring outside of the laboratory, stand out as probable evidence for the reality of precognition. They shed new light on the terms *fate* and *destiny*. Was the *Titanic* fated to sink? Was Patty Hearst fated to be kidnapped? How can we explain these events?

Every event is predetermined in that it is caused by previous events. What changes is the time element. Take, for example, the toss of a coin. At some point during the "flip" the outcome becomes irreversible. For each coin toss, however, the time factor may vary.

I remember flipping baseball cards as a youth. The way the card was held and the way the wrist was flicked contributed significantly to the way it landed. With practice, one could flip a card and gain a tremendous amount of certainty as to how it would land. So, as kids, we would challenge each other to see who was the best at it. In this instance, the time factor varied along with wrist action and stiffness of the card. The more these variables were controlled, the more likely the result. However, one could never get what one wanted on every throw; otherwise, there would be no game. Nevertheless, at some point during the flight of each card, the end result, of the picture landing faceup or facedown (or leaning on a wall), became predetermined.

In the case of the building of the *Titanic,* there was a probable future that it would hit an iceberg and sink. The problem with this case, however, is that this probability seems to have existed years before the ship was even designed and built!

Perhaps a more personal example of this kind of prediction may shed some light on the process. In 1989 I wrote a screenplay entitled *Hail to the Chief,* about a third-party candidate who runs for the presidency. My thinking was that if a candidate were famous enough, he or she could usurp the two-party system. Examples included the actor Robert Redford; Ted Turner, winner of the America's Cup and entrepreneur; and Lee Iacocca, chairman of Chrysler Corporation.

I then set up the premise that there were horrible yet difficult to prove scandals involving the leading Democratic and Republican contenders. This was the impetus to spark the third-party candidate to enter the race. I chose a newscaster, someone like the late Peter Jennings or Tom Brokaw. To add to the mix, I included a fourth major contender, a televangelist. I made the Democrat a female senator and the Republican an African-American general.

At the time, I vaguely knew about Colin Powell, who became head of the Joint Chiefs of Staff in September 1989. I may have written the screenplay before his advancement, but he was not well known at that time. He had no party affiliation, and his visibility really came into being a few years later during the Gulf War. Concerning the televangelist, I modeled the character in certain ways on Pat Robertson, who did, indeed, run for the presidency. I felt at the time that there could be a younger version of him on the horizon.

In any event, now we jump eighteen years. The leading contenders for the 2008 presidency at this moment, December 2007, include three Republicans: Rudolph Giuliani, former mayor of New York; Mitt Romney, former governor of Massachusetts; and Mike Huckabee, governor of Arkansas and former televangelist. The three Democrats are Barak Obama, a first-term senator from Illinois; John Edwards, a former senator; and Hillary Clinton, former first lady and second-term senator from New York.

Although a major newscaster has not entered the race, there has been a call for Lou Dobbs, a news announcer from CNN, to enter the race. So my similarities include an African American, a televangelist, and a female senator, and a distinct possibility of a newscaster running as well. In the screenplay, the African American was a Republican, whereas Obama is a Democrat. The televangelist was a renegade Republican who ran as an Independent, whereas Huckabee is just a Republican, and the female Democratic senator is a match. The main premise, however, is missing: the idea that a third-party candidate could usurp the entire three-party system. There is, however, the distinct possibility that Michael Bloomberg, a mayor from New York, who succeeded Giuliani, may enter the race as a third-party candidate. *Newsweek* ran a front-page story on Bloomberg with precisely this possibility.

Are my hits close enough to the mark to be "synchronistic?" I had an African American, a woman, a televangelist, and a call from the public for a specific newscaster to enter the race. Also, the idea of a significant third-party candidate entering is a theme of the 2008 presidential race. Certainly when I wrote the piece, nearly twenty years ago, I thought I was dealing with inevitability. It was inevitable that women and racial minorities would eventually emerge. I also felt that it was inevitable that at some point, a third-party candidate would emerge. This has occurred in other countries such as Hungary, where the famous pianist Ignace Paderewski became prime minister of Poland. The well-known poet Vaclav Havel became head of the Czech Republic, and, in America, Jesse Ventura, a pro wrestler running as an Independent, became governor of Minnesota.

I think the key here is the ability to determine what is inevitable.

Synchronicity, or the mystical element, enters when something amazing coincides. Whether or not this occurs, in this instance, is subject to debate.

Inherent in any discussion of precognition is the problem of the structure of time. If "pure precognition" actually occurs, there is a presupposition that at least some events are determined outside the normal spacetime realm.

THE SPACETIME MIND CONTINUUM

If we are to come to a true understanding of the nature of the laws of the universe and such things as the structure of time, there has to be a reconciliation between physics and metaphysics. The reality of the existence of our minds must be integrated with the structure of the spacetime continuum. However, if we start with the hypothesis that the mind must be part of the structure of the universe and reflect upon the four-dimensional model for spacetime,[7] we encounter numerous difficulties. As the theory is attempting to integrate the psyche into the so-called physical universe, the final conclusions need to be looked at as a gestalt. The four-dimensional model is basically one that allows the three physical dimensions of height, width, and depth to exist in the fourth dimension, time.

Fig. 10.5. The Persistence of Memory, *Salvador Dali*

We could, perhaps, concede that the past is part of the four-dimensional present, as the present is the product of the past. But the future is (for the most part) not part of the present. This is by definition.

10.6. The three-dimensional cube exists in time, the fourth dimension.
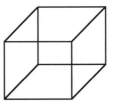

A brief reflection upon this cube in time with regard to past, present, and future could be diagrammed as follows:

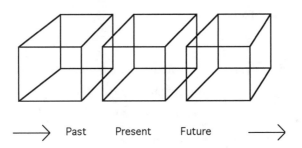

⟶ Past　Present　Future ⟶

Fig. 10.7. This diagram attempts to depict the echoes of the past and future of the three-dimensional cube existing in the fourth dimension. The solid cube is analogous to a standing waveform in a sea of vibrations, as the structure of the cube persists over time.

If we speculate about where the three-dimensional cube will be in the future (that is, in time, or the fourth dimension), we have already projected the imagined cube into any of a series of possible "places." This simple ability to postulate about tomorrow cannot be explained in an obvious way by the present four-dimensional model. By using our minds to project into the future, or to reflect upon the past, we enter and utilize a realm that is not part of the physical spacetime continuum but instead travels with it.

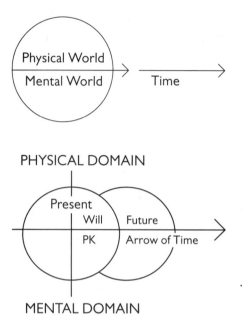

Fig. 10.8. The present three-dimensional world can proceed into the future in a variety of ways. These exist as potentialities in the second dimension of time, which appear here in the mental world.

Fig. 10.9. The will operates as the present becomes the future, that is, when a mental even is converted into a phsyical one. Psychokinesis operates in the same or similar fashion.

Experiences from the immediate past blend with the present, as do experiences from the immediate future. The circles in figure 10.10 are bisected; the top half depicts the progression of physical events and the bottom half represents accompanying mental events. Note also the implied spiral nature of onward progressing time. The mind, however, is able to permeate both realms. If we include the past, present, and future, we come up with the following diagram:

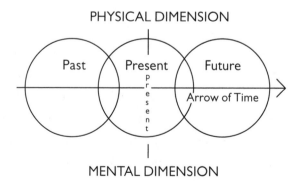

Fig. 10.10. The physical and mental world existing in the past, present, and future

A similar scheme can be used to explain a number of terms (see fig. 10.11). Telepathy "is the awareness of thoughts of another person";[8] precognition involves knowledge of a future event during the present moment;[9] retrocognition involves the obtaining of information from the past. Rhine defines *clairvoyance* as "the awareness of objects or objective events without the use of the senses."[10] For the sake of this discussion, clairvoyance will also be defined as a broad term encompassing telepathy, precognition, and retrocognition.

From this point of view, just as cardinals, sparrows, and pelicans are different types of birds, telepathy, precognition, and retrocognition are different forms of clairvoyance. By defining *clairvoyance* as simply paranormal information retrieval, all forms of extrasensory perception become unified. Clairvoyance also involves obtaining information from inert matter (psychometry) and from the departed (mediumship).

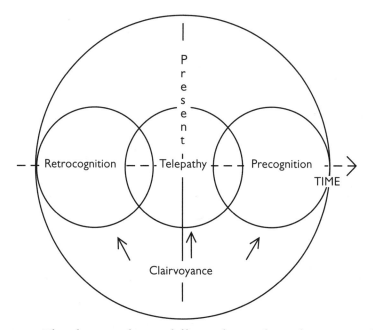

Fig. 10.11. This diagram depicts different forms of psi along a time line. Retrocognition involves looking into the past, telepathy occurs in the present, and precognition involves receiving information from the future. All forms of ESP that involve information retrieved can be conceived of as forms of clairvoyance.

A more complete drawing might look as follows in figure 10.12. Reading from left to right, we note the progression of time. Determinism, a force from the past, creates the chain of cause and effect; however, events emanating from the past are also influenced by future goals, that is, the teleological component.

The three circles represent past, present, and future and the accompanying forms of cognition. In the physical domain, psychometry would involve obtaining information (psychically) impressed into the vibratory state of matter. Normal perception from the five senses occurs in the present, and predictions based upon imagination combined with knowledge of current physical determinants involve the sphere of the

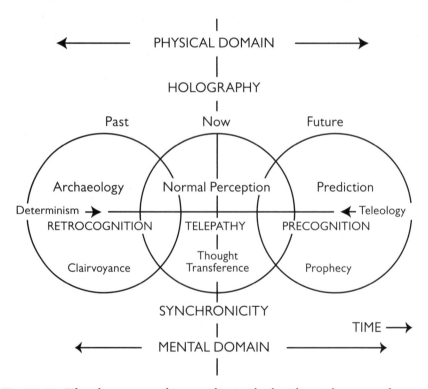

Fig. 10.12. This diagram combines and extends the ideas inherent in figures 10.8, 10.10, and 10.11. It attempts to depict the physical and mental correlates to retrocognition, telepathy, and precognition. Normal sense perception is represented in the top sector of each circle, and paired below with its corresponding paranormal ability.

future. Beneath prediction are precognition and prophecy, both involving paranormal information deriving from "nonphysical" dimensions of the future. Telepathy, or thought transference, involves psychic ability in the present, and mediumship involves obtaining information from people who lived in the past.

J. W. DUNNE: AN EXPERIMENT IN TIME

One name that stands out in relationship to theories related to precognition and the structure of time is that of J. W. Dunne. His famous work *An Experiment in Time,* published in 1929, contained the analysis of his personal precognitive experiences along with his model for a multidimensional universe. His ideas continued to expand, and in 1934 he wrote a sequel entitled *Serial Universe.*

Dunne's initial interest in this topic was sparked by striking dreams that he had about two events, each of which occurred before the newspaper reports of the events were actually published. "Indeed, the more I thought of the two episodes, the clearer it became that, in each case, the dream had been precisely the sort of things I might have expected to have experienced after reading the printed report."[11] Dunne considered the possibility of a telepathic linkage with the journalists, stating that certain precognitive events could be explained by either clairvoyance or thought transference. But other instances, he concluded, were definitely "precognition." For example, in a dream he saw a horse escaping from a fenced enclosure and charging madly at him. The following day the actual event took place. Dunne pointedly writes, "[S]ome particulars of the dream were accurate . . . like wooden stairs that lie at the end of the path the horse was charging on."[12] Other facts, however, were wrong, as the horse was in a field that was different from the one in the dream. Dunne concludes that the dream was displaced in time. He thus seeks to analyze the structure of time and thereby develop a model of spacetime that is more in accord with observed experience.

In the introduction to *Serial Universe,* Dunne writes that "new facts at first appear unintelligible. . . . The empiricist puts the facts before reason whereas the rationalist puts reason before the facts." Dunne suggests that if "the Universe is a product of mind . . . then it will ultimately illustrate mind's axiom."[13] From this premise, Dunne proceeds to describe an elaborate universe that puts much emphasis on predetermination. The well-known writer on the nature of time G. J. Whitrow traces out the roots of Dunne's thinking. He believes Dunne's precognitive episodes could be looked at as "pre-presentations."[14]

> To account for these, and other alleged phenomena of precognition, which may be defined as non-inferential knowledge of future events, he formulated a theory of "serial, or multidimensional time" (Dunne, 1934). This was an ingenious development of the hypothesis originally put forward by C. H. Hinton (1887) that the world is a four-dimensional spatial manifold and particles are "threads" in it. Human beings are only perceptually aware at any instant of a three-dimensional cross-section, so that in effect they seem to be "traveling" along the fourth dimension. This "traveling" is, however, merely a progressive transference of awareness to one cross section after another, producing the illusion that there is a three-dimensional world enduring in time and that parts of it are in motion. According to this hypothesis the world is static and the illusion of time arises from the continual change of the observer's attention. Dunne, however, realized that the continual transfer of attention is itself a temporal process and so could not produce the time necessary for its own occurrence. To account for this time, he postulated that the manifold has a fifth spatial dimension and that a second consciousness "travels" along it. However, as the same difficulty now breaks out all over again, he was obliged to postulate an infinite number of extra dimensions and a corresponding number of observers. Precognition is possible in such a world because time is unreal.[15]

Regarding Dunne's work, Saltmarsh concluded that the infinite time dimensions were untenable and unnecessary because, among other reasons, there could be no last dimension. Saltmarsh is also particularly critical of Dunne's static concept of the dynamics of life's events: "It seems to me that the proper starting place for any theorizing is change and not time at all. . . . The fundamental analysis is incorrect. Time does not flow over a static history at a certain rate. History is that, which is left behind by the onward surge of change."[16]

Saltmarsh suggests that precognition is commonplace. "Coming events cast their shadows . . . such as in the rumbling of a train before it passes."[17] Saltmarsh writes that much of precognition involves prediction. "What we want to discover is whether there are any cases where knowledge of the future is not based upon inference from knowledge of the past and present."[18]

This search for "pure precognition" presupposes another dimension to spacetime, which would allow a realm for these upcoming episodes to exist in the present. Obviously, if events can be foreseen, then this other dimension has a mental component. Future events existing in this realm, however, are "not inexorably fixed but are capable of being modified by deliberate action beforehand."[19]

Saltmarsh suggests "that the crux of the problem is time"[20] and change, for a "world with no change would be timeless."[21] His solution is to allow time to be two-dimensional. Our conscious self is aware of the present, whereas the subliminal self has a "larger" view, which encompasses more of the future.

Precognition, he says, involves the "subliminal mind," a subconscious stratum that may receive information concerning future events (existing already in this underview of the present), and transmit this forthcoming event to the conscious mind by means of symbols.[22] "This theory therefore explains precognition by denying it; that is to say, it suggests that, which appears as a case of supernormal non-inferential foreknowledge, is only a fragment of the knowledge of the present of the subliminal mind, which is thrust up to surface consciousness."[23]

This is to say that our surface self has only a limited perception of what the real present is, because the real present also encompasses conscious and unconscious desires and plans of individuals and the masses. For instance, in a subconscious way, the bodybuilder-turned-actor Arnold Schwarzenegger over many years cultivated such an energy that he made it possible to become governor of the state of California. In conscious and unconscious ways, many people in that state were willing to vote for him before he actually announced his candidacy. In some mental virtual space, Schwarzenegger became governor long before the election because he was already governor in the minds of a majority of people. Saltmarsh concludes, however, that subliminal knowledge of an event that already exists in the second dimension of time does not mean that the event must occur, as this realm perceives events that are "to some extent plastic."[24]

To extend this example further, Schwarzenegger is also already president of the United States in one of the "halls of future probabilities." However, he carries one huge stumbling block and that is a restriction in the Constitution that disallows foreign-born citizens to take the highest office. The second dimension of time does not necessarily manifest itself in the first dimension of actuality. Even though it may exist as a potentiality, that potentiality may only dwell in the land of the unconscious in the same way that most of the dreams that we have throughout our life never pierce the world of conscious reality. To quote Carl Jung, "The unconscious really is unconscious!"

FIVE-DIMENSIONAL UNIVERSE

Saltmarsh is the rare theorist who has taken into account the structure of consciousness in relationship to the structure of time, although Ouspensky's writings suggest such a linkage. Saltmarsh's "subliminal self" (that is, unconscious mind), which to some extent extends beyond the present, and Rampa's "hall of future probabilities"[25] are similar to

Dunne's ideas of the term "second observer" (that is, another aspect of the observing ego).

Dunne suggests that this second observer watches the normal self living in three dimensions of space and one of time as a four-dimensional space. This idea is intriguing, because we generally accept the fact that the physical world has only three dimensions. Thus, following this general line of thinking, if we include the realm of mind (or inner space), a fourth spatial dimension, the new model for the space/time/mind continuum would be five dimensional:

- Three spatial dimensions
- One dimension for inner space, such as the mind
- One dimension for time, that is, the present

The fourth spatial dimension, which we can call inner space, could also be looked at as the first dimension of time, where future probabilities exist as potentialities. The next dimension, the one we generally equate with time, is the actual present, where all potentialities collapse into the realm of the ongoing actuality of the "Now." The realm of inner space can also be called counterspace, mind, or hyperspace (all these terms being essentially interchangeable).

The Necker cube, seen in figure 10.13, is one way to demonstrate the existence of a fourth spatial dimension. When you stare at the cube, note how the shifting of perspective can occur only in the mind. The drawing on the paper has not changed. This shift takes place in inner space.

> This transformation between three-dimensional congruent but non-superimposable counterparts (for example, a pair of gloves) can only take place in manifold (space) that is at least four-dimensional. . . . Since we do not see more than three dimensions (height, width, and breadth), how can we identify this extra dimension? . . . We should regard it as an imaginary time dimension.[26]

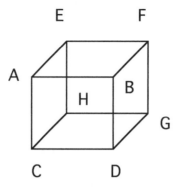

Fig. 10.13. When staring at the cube, sometimes ABCD is in the foreground and sometimes EFGH is. This shift does not occur in physical space.

The present-day four-dimensional spacetime model has left no "space" for mind. By introducing the idea of inner space as an actual fourth "space," time then becomes the fifth dimension. The inner realm, which can also be called hyperspace, or even virtual space, is a place where the future exists as a series of probabilities. In actuality, *all* mental phenomena would also exist in this realm, as it encompasses the entire Freudian/Jungian model of a conscious, preconscious, unconscious, and collective unconscious (outlined in depth in *Inward*

Fig. 10.14. P. D. Ouspensky, circa 1935

Journey). According to both Freud and Jung, the unconscious, which is barely known by the conscious, *thinks*. It has its own set of laws; it is infinite in size and interpenetrates all mental domains, including the so-called astral planes. This realm not only influences the conscious, but it also senses future events or future probable events because it has, as Saltmarsh suggests, "a wider view of reality."

OUSPENSKY, HYPERSPACE, AND THE FOURTH DIMENSION

Excessive complexity in any construction is always the result of something having been omitted or wrongly taken at the outset.

P. D. OUSPENSKY

Ouspensky describes how the first three physical dimensions are created.[27] Each evolves from the other in the same way. They start from a point, which is the zero dimension, and evolve into a line, plane, and then cube:

Fig. 10.15. A line can be viewed as a point moving through space.

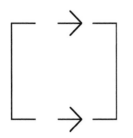

Fig. 10.16. A plane can be viewed as a line moving perpendicular to itself.

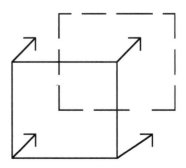

Fig. 10.17. A plane becomes a cube by moving perpendicular to itself.

Further, a cross-section of a line is a point, a cross-section of a plane is a line, and a cross-section of a cube is a plane. Therefore, a cube can be seen as a cross-section of a four-dimensional object. Thus the existence of a cube presupposes the existence of a fourth spatial dimension. Although this extra dimension does not exist in three-dimensional space, it can be drawn. It has been called a hypercube.[28]

When a line moves perpendicular to itself for a length of A to produce a plane, the square formed becomes A^2 (fig. 10.16). The cube can be measured as A^3 and therefore the hypercube has dimensions of A^4 (fig. 10.18). Since the three spatial perpendiculars account for the physical world, a figure with dimensions A^4 would have a "perpendicular unknown to us in our space."[29] "Admitting the existence of the fourth dimension, we must recognize in the same way that if there are four dimensions, the real body of three dimensions cannot exist. . . . [For what we are seeing is] the cross-section of a body that is really existing in four dimensions."[30]

Ouspensky suggests that our eye sees in three dimensions but we perceive in four dimensions. "From the fourth dimension it must be possible to see the cube from all its sides at once from within, as though

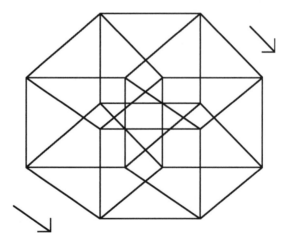

Fig. 10.18. The hypercube. A cross-section of this four dimensional object is a three-dimensional cube. Can you see the cube at the top left moving down and "becoming" the cube at the bottom right?

from its center."[31] Just as the cube exists in a dimension perpendicular to the plane, the fourth dimension is at right angles to the other three in a perpendicular unknown to the physical world. "Vision in the fourth dimension must be effected without the help of eyes . . . (and) must be something quite different from ordinary vision."[32]

This concept of changing dimensions by moving "perpendicular to itself" is related to the concept of orthorotation, or angular momentum. Since the Earth spins on its own axis, we can see that, as it is moving "perpendicular to itself" (that is, the equator in relationship to the north-south axis), it may also be continually changing its dimensional properties. Keep in mind that the property of spin is also directly related to the imaginary number i cited above, and thus the realm of mind. Such gyroscopic properties are also associated with the mysterious ability of spinning objects to set up their own coordinates in space regardless of the spin and position of the Earth.

Ouspensky's conclusion is that the fourth dimension is the dimension of mind. We are defining it as synonymous with the term *hyperspace* as well as "inner" space. Therefore, by its very nature, it transcends normal three-dimensional space. It is cross-sectional to three-dimensional space.

> The psychic, as opposed to the physical or the three-dimensional, is very similar to what should exist in the fourth dimension, and we have every right to say that thought moves along the fourth dimension.
>
> No obstacles or distances exist for it. It penetrates impenetrable objects, visualizes the structure of atoms, calculates the chemical composition of stars, studies life on the bottom of the ocean, the customs and institution of a race that disappeared tens of thousands of years ago. . . No walls, no physical conditions restrain our fantasy or imagination. . . . The greater part of one's being exists in this fourth dimension, but we are unconscious of this greater part of ourselves.[33]

One sentence is worth repeating: "The greater part of one's being exists in this fourth dimension." Normal cognition and psychic phenomena, such things as thought transference, synchronicity, and precognition, may be overt manifestations of the four-dimensional world expressing itself in three-dimensional reality. This concept of a four-dimensional space is compatible with any theory that attempts to place the mind within the spacetime structure. The mind, existing in the fourth dimension, has the ability to transcend three-dimensional spacetime, by definition. This realm, "perpendicular to the other dimensions," exists in "inner space."

COUNTERSPACE: THE PRIMARY VIRTUAL REALITY

Rudolf Steiner has called this realm "counterspace."[34] Whereas the three physical spatial dimensions are projected outward into infinity, counterspace proceeds inward. There are no ordinary geometrical points in counterspace, as each point in this realm opens up to a new infinity. A physical prototype for this theory can be found on the Internet. For instance, consider the hyperlink. One can go from virtual world to virtual world to virtual world via this method, and each hyperlink is a potential infinity unto itself. This vast new arena, called the World Wide Web, can easily be broken down into a binary system of zeros and ones, as computer chips listen only to whether the current is on or off. Similarly, the neurons in the brain either fire or they don't; the brain itself can be broken down into an electronic binary system comparable to the one we have created with the Internet. Both provide a way for us to understand how the All can emanate from the One and its negation.

Another prototype of counterspace is demonstrated by movie history. One evening in 2003 I was watching, with my wife, Lois, a new Western starring Tom Selleck and Isabella Rossellini. In this film, we both commented that Isabella greatly resembled her mother, Ingrid

Bergman, a woman I have seen only in movies. Some of the scenes from the Western called to mind the 1940s masterpiece *Casablanca,* arguably the greatest movie ever made, starring Ingrid Bergman and Humphrey Bogart. This WWII–era watershed achievement exists in its own virtual reality in the minds of millions upon millions of people. From that film, fictional future possibilities for the Bogart/Bergman couple can be projected. The realm where this takes place exists somewhere in Lobsang Rampa's hall of future possibilities, a different, more complex dimension from the physical plane. Creepily, with the new advances in digital special effects, a completely realistic movie could be created with these characters at some point in the future. Other potential movies that live on as possibilities in the minds of many people are sequels to *The Maltese Falcon* and *Body Heat,* and another buddy film with Paul Newman and Robert Redford.

The fictional lives of such characters as Rick, Ilsa and Victor Laszlo live on in a very real virtual place, and their lives—like those of Tom Sawyer, Huck Finn, Sherlock Holmes, Moriarty, Captain Ahab, the white whale, Bugs Bunny, Miss Piggy, James Bond, Spiderman, the HAL computer, Spock, ALF, and Mickey Mouse—continue on into the future although these fictional characters never existed in the physical world.

In a similar way, many actual people take on a comparable mythic aspect. Such individuals as Abe Lincoln, Teddy Roosevelt, Adolf Hitler, Genghis Khan, Marilyn Monroe, Elvis Presley, Frank Sinatra, JFK, Jesus, Buddha, and Mohammed come to mind. In the case of Frank Sinatra and his "Rat Pack," which includes his pals Sammy Davis Jr., Shirley MacLaine, and Dean Martin, their songs and skits have been preserved on tape so that future generations can experience them for many years to come. This is real energy that generates emotion in the minds of the listeners, who in turn create their own energies, causing spin-off virtual realms such as the *Ocean's 11–13* remakes recently produced by George Clooney and his Rat Pack: Brad Pitt, Matt Damon, Don Cheadle, Carl Reiner, Andy Garcia, and Julia Roberts.

The realm of mind is, or operates in, an extra spatial dimension. But—just as a person can take an idea (such as a blueprint for a house) and manifest it on the physical plane (that is, build the house)— properties of inner space involve the ability to become manifest in outer (physical) space or remain energetic in various conscious or subconscious realms. Our brains and bodies house the capability of transforming mental ideas into concrete physical constructs. Thus, we lie on the interface of the third and fourth dimensions.

Another feature of inner space is that it permeates the other three spaces. In this sense, the realm of mental inner space (hyperspace, counterspace, or mind) can be seen to be compatible with Freud's idea of the libido, a biological force that transforms mental energy into physical energy (or vice versa). It is simply the mental plane of existence, which lies beneath the physical spacetime environment. It is structured hierarchically, its first levels lying where our "normal" human minds are (in an everyday state of consciousness) while its deeper strata are connected to the Freudian unconscious, memories, dreams, and so on, to Jungian archetypes existing in the collective psyche, and to corresponding neuro-electric counterparts.

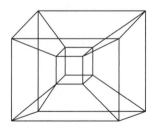

Fig. 10.19. Another type of hypercube. This one could be visualized dynamically as a smaller Necker cube "becoming" a larger cube, or vice versa.

SUMMARY

A full description of our environment, called by the physicist the spacetime continuum, generally ignores the realm of mind. Phenomena such as synchronicity, telepathy, and precognition are also ignored. But, as shown here, a realm for the mind does exist and there is abundant evidence that synchronistic events and precognitive experiences are actual phenomena. The fact that humans can project into the future presupposes a dimension of time not truly handled by the current four-dimensional model. Just as there is an outer physical world, there is also an inner mental world; by viewing these concepts spatially, a five-dimensional mind/space/time model has been developed. The extra dimension extends inward into inner space, called the fourth dimension by Ouspensky (1931/1977), hyperspace by Rucker (1977), and counterspace by Steiner and Whicher (1971). It is also the realm of the mind and is structured in such a way that it is compatible with both the Freudian and the Jungian models.

Synchronistic phenomena can be explained by the utilization of a fourth spatial dimension for the simple reason that all physical objects are connected in special ways (such as by meaning) in hyperspace, the realm of existence that can become manifest in the outer three dimensions (in "time").

The hyperspatial realm, since it proceeds inwardly, can also contain an infinite number of possible futures existing as probabilities. Some of the future probabilities that actually manifest later on the physical plane can be recognized before they occur simply because they exist as potentialities in this space, which also houses our minds. The realm that contains the seeds of all future probabilities can also be thought of as the second dimension of time.[35] However, these probable futures do not necessarily have to become manifest on the physical plane.

A brief reflection on the structure of everyday events reveals that all planned human activity is initiated by thought. The realm for plans, goals, and fantasies exists in this dimension called mind, a realm that can be looked at from a psychoanalytic perspective, but can also be conceived of as the product of electronic brain function. By giving this realm a "space," it can be linked to the physical spacetime continuum.

In order to explain bizarre coincidences such as the *Titan/Titanic* episode, we can speculate that the future already existed in the realm of inner or hyperspace as a probable future. In such coincidental events, various factors such as initiating archetypes, telepathic and other psychic activity, and even chance all played a part in their manifestation. Future probabilities play out in the multidimensional realm of mind, and sometimes they also manifest on the plane of physical reality.

The realm of inner space is beneath the surface of physical reality. It is structured in a lawful way, but quite different from three-dimensional space. One of its principles is compatible with the holographic model in that somehow the totality of hyperspace permeates all levels of physical space at all times at once. One cannot view this statement with a three-dimensional logic. For want of a better term, we could call the inner realm the mind of God.

In the case of large-order synchronistic experiences—precognition being one form of synchronicity—the understanding that a substratum of mind permeates the spacetime continuum provides Jung's initiating archetypes a space in which to operate. Besides each individual mind, there is a mental counterpart of the very structure of our spacetime environment. The recognition of this hyperspatial realm (as a mental domain) is the next essential step toward a comprehensive theory of the structure of the universe.

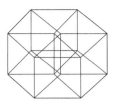

11
Six-Dimensional Universe

Higher dimensional geometry may be the ultimate source of unity in the universe.

Michio Kaku,
Hyperspace, 1994

Ouspensky notes that since time is the measure of the movement of the Earth through space, it takes the form of a spiral. He writes that it takes three coordinates to describe the spiral: duration, velocity, and direction; time thus has three dimensions. This idea would also make the threefold corkscrew structure of time symmetrical to three-dimensional space. Ouspensky's six-dimensional model is basically the same as the four-dimensional spacetime model in physics with the addition of two dimensions for time.

Simply stated, the fourth dimension of space would be the line of historical time. It is also the first dimension of time, comparable to that established by Einstein. The fifth dimension of space is also the second dimension of time. This realm houses all past and future probabilities (see fig. 11.1 on page 238).

This dimension is somewhat similar to Dunne's and Saltmarsh's ideas on two-dimensional time, as well as Everett's "many worlds interpretation" of quantum mechanics (1973), which allows for all possibilities to exist simultaneously. All probable futures—shown in the diagram as being perpendicular to the fourth dimension—exist in the fifth spatial dimension (or second dimension of time) forever.

↑	1800	1900	2000	2100	↑	
Line of Historical Time — Past →			Present →		Future → →	
↓	1800	1900	2000	2100	↓	
↓	1800	1900	2000	2100	↓	
↓	1800	1900	2000	2100	↓	

Fig. 11.1. This diagram, adapted after Ouspensky (1931/1971), depicts the relationship between the first dimension of time (the line of history) and the second dimension of time, existing as a realm of probable futures (for all times) running perpendicular to the first dimension.[1]

Each future moment in time exists as an infinite number of possibilities, although, technically, Ouspensky states the number of future possibilities really cannot be infinite because some probabilities are really impossibilities. For instance, Franklin Roosevelt is not going to run for president again, and Steven Spielberg will never ride a unicycle on Mount Everest. That said, the fifth dimension of space (the second dimension of time) is compatible with theories about precognition, as all future probabilities exist in this realm. Therefore they can be cognized before they become manifested in the fourth dimension, that is, the present. Most future possibilities obviously never materialize. Precognitive experiences involve seeing probabilities, that is, events that are more or less likely to occur.

The probabilities that have never manifested still exist in realms of "perpetual nows," parallel times existing in the fifth, or hyperspatial, dimension (the second dimension of time). They still, to some extent, influence the line of actualization (that is, the line of the fourth dimension).[2] For example, suppose a person narrowly misses an accident while driving on a particular street. The thought of "what could have been" (which exists in the fifth dimension) may cause that person thereafter to always drive slowly on that street even though no accident occurred. In this instance, we can see how a probable future from the past still influences present and future events.

But let's think more deeply about this. This second time dimension, or fifth dimension of space, must be arranged hierarchically. All past and future probabilities in the minds of the living would have a lot more energy than those same dimensions for the departed. For instance, many people would like to see Al Gore become president. He could run six or ten years from now, if his health remains. I had a friend who wanted to become a lawyer, but took another path when he was in his early twenties. When he hit forty, he went to law school, so his past probable future took about twenty years to manifest. But what about the probable futures of people who lived one hundred, two hundred, or five hundred years ago? Consider also the ethnic hatreds that continue to manifest even hundreds of years after the assumed original insult.

If we wanted to delineate the subspaces of this second time dimension we would have to consider not only the practical aspects existing in the minds of the living as compared to the dead, but also possible versus impossible futures, and also whether or not is there an artifact attached to the physical dimension. For example, suppose an Incan city of gold has been buried for seven hundred years and then is discovered. If the Peruvian inhabitants of that region prosper because of incoming tourists, and so on, in some way the future plans of that ancient city and the people of that city were able to be made manifest because their departed probable futures left some trace in the physical dimension.

In terms of the concept of precognition, the main consideration for this dimension is probable futures in the minds of the living. For example, let us consider an upcoming presidential election. If subconsciously, even without their *conscious* knowledge, people's minds are already made up for a particular candidate, then to that extent, that candidate is already president in one of the spaces of this second dimension of time. Thus, a dream of that man or woman as president is precognitive, but only potentially so, that is, if that possible future is the one that becomes manifest.

The sixth dimension of space (or third dimension of time) would be probable futures "off" the probable futures of the fifth dimension. They would therefore encompass all and everything.

This realm is related to omniscience, eternity, and the ever-advancing present moment, the ongoing *Now*. In its full potentiality, it is incomprehensible to the fifth dimension. The present moment that you are in at this very instant, this *Now*, is its own special space. If we refer to the chart on page 241, we note a dimension for hyperspace and a second and third dimension for time.

THE PHYSICS OF HIGHER DIMENSIONS

The mind occupies time, but does not occupy 3-space.

TOM BEARDEN

In order to fully understand the progression from one dimension to another, a clearer understanding of the word *infinity* is needed. Ouspensky points out that *infinity* is rarely utilized correctly. One of the problems in Einstein's universe comes from a misunderstanding of this term. "Infinity has a definite meaning only in mathematics. In geometry infinity needs to be defined, and still more does it need to be defined in physics. . . . And the establishment of different meanings of infinity solves a number of otherwise insolvable problems."[3]

Infinity for a line, he writes, can have two possibilities: "A line continued into infinity, or a square. . . . What is infinity for a square? An infinite plane, or a cube."[4] Thus, the term retains its original meaning, "but to it there is added another, the concept of infinity as a plane resulting from the motion of the line in a direction perpendicular to itself. . . . For every figure of a given number of dimensions,

Dimension		Einstein	Ouspensky	Saltmarsh/Dunne
SPACE				
Point	$0 = $ zero	Same	Same	Same
Line	$1 = A^1$	Same	Same	Same
Plane	$2 = A^2$	Same	Same	Same
Cube	$3 = A^3$	Same	Same	Same
Hyperspace	$4 = A^4$	-----	4th dimension, realm of mind. Similar to Steiner's idea of counterspace, and O's 2nd time dimension.	-----
TIME				
TIME Earth's dynamic position in space in relation to the sun.	4	Relative to observer. Appears as imaginary number $\sqrt{-1}$ equivalent to 3-D space, à la Minkowski. Used to locate position of moving objects.	Line of historical time, 4th D of space and/or 1st D of time.	First dimension of time is an illusion for Dunne. Observer moves along predetermined path. For Saltmarsh, normal cognitions of conscious self.
	5	-----	5th D of space and 2nd D of time, realm of probable futures.	Observer sees normal 3-D space and 1-D time spatially; however, time is multi-dimensional for Dunne. Perceives wider view of future by subliminal self. A 2nd D of time for Saltmarsh.
	6	-----	3rd D of time. Eternity, the NOW, also omniscience. The All beyond human understanding.	-----

Fig. 11.2. Space/time/mind models

infinity is a figure of the given number of dimensions plus one. . . . Incommensurability [between dimensions] creates infinity."[5]

> The problem of infinity is encountered in physics with the use of the various Lorentz transformation equations that relate length, time, or mass to velocities approaching light speed. According to these equations, nothing can travel as fast as the speed of light, because if it did, it would obtain infinite mass. Ouspensky writes that although the physicists won't admit it, "when velocity reaches infinity [that is, the speed of light], it becomes *something else*. But physicists certainly could not surrender at once and admit that velocity can cease to be velocity and become something else. What did they stumble upon? They stumbled upon an instance of infinity. The velocity of light is infinity as compared with all the velocities that can be observed or created experimentally. And, as such, it cannot be increased. In actual fact, it ceases to be velocity and becomes an extension. A ray of light possesses an additional dimension as compared with any object moving with terrestrial velocities."[6]

Ouspensky suggests that Einstein never solved this problem of infinity for the simple reason that light speed is incommensurable with velocities of physical objects. The implication now becomes obvious. The velocity of light demarcates the boundary to the next higher dimension. This hyperdimension is infinity for physical objects in the same way that a line is infinity for a point. And this realm, as has been suggested numerous times, is commensurate with that of the mental sphere.

Ouspensky's six-dimensional universe could be mapped out as a simple graph where a ninety-degree turn and movement out creates the next dimension and each dimension is infinite to the one before it. The simple flat graph below is actually a two-dimensional representation of a multidimensional spiralized concept.

The first cause, or the impetus of the universe, can be seen as propelling time forward, but future goals inherent in various structures

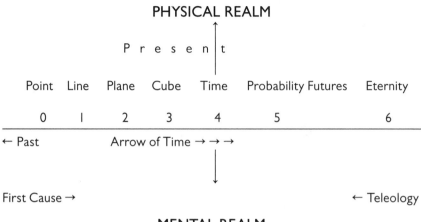

Fig. 11.3. This diagram depicts the nature of time, starting with a simple zero dimension and progressing upward through Ouspensky's three dimensions of space and three of time. Note that the first four dimensions comprise "the present." Therefore, past and future are to the left and right of this four-dimensional spacetime realm.

(such as DNA) also direct the flow of time, and in that sense, this energy travels from a future-expected point back to the present. In hyper-, inner-, or counterspace, every point therein is a window to a new infinity. Time is transcended and new mental worlds become available. To completely track this realm, we can extend the Freud/Jungian model to allow for an essentially infinite number of mental domains or dimensions, many of which are barely known or not known by the conscious mind.

SPIN

Clearly there is a mystery associated with spin, gravity, ether, and the idea of dimensions. Perhaps if we study the gyroscope, we may gain some understanding about where all this is leading. As it turns out, the qualities of a gyroscope are most peculiar because a force exerted on a spinning gyroscope will be converted to one at 90 degrees to the initial direction of the force. Note the antigravitational effect created in the diagram below:

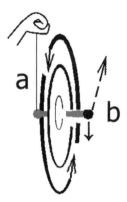

Fig. 11.4. A gyroscope

From this diagram, we see two opposing forces. The momentum of the spin of the gyroscope, the dotted upward arrow, opposes gravity, which is seen as a downward arrow at "b." So, instead of the gyroscope simply flopping down, it *precesses,* that is, it rotates around at "a" in a circle. Thus we see that the gravitational pull will be converted to a force perpendicular to the vertical. Another property of a gyroscope is that it sets up its own coordinates in space regardless of the motion of the Earth. That is why the gyroscope can be used as a compass.[7]

The question remains as to how these curious properties play themselves out in the elementary particles, particularly when we consider that they are spinning at speeds in excess of the speed of light. One thing that may be occurring is the combination of two spins at the same time, one in the direction of the spin of the elementary particle and the other somehow linked to the process of precessing, which would be at right angles to the primary spin. If we consider what Ouspensky is saying, namely that the change of dimensions is associated with an orthorotation and movement out, all of this suggests that elementary particles interface dimensions from a prephysical etheric one out to the physical one. Obviously, one of the principles of these spinning electric charges is to manifest themselves in the form of atoms, the elements, and thus matter.

THE MYSTERY OF SPACE

Equations demonstrating the arithmetic and algebraic existence of $\sqrt{-1}$ (such as $X^2 + 1 = 0$) had been well known at least since the Italian Renaissance, but it took about a half a millennium more for these so-called imaginary numbers to become accepted by men at large, when it was found that without using $\sqrt{-1}$ satisfactorily, simple representations of electromagnetic phenomena and effects were not possible.

CHARLES MUSÈS

As discussed in chapter 4, the use of the imaginary number $i = \sqrt{-1}$ by Minkowski, Einstein, and Dirac enabled scientists to better understand both the nature of spacetime and the structure of the atom. Paradoxically, imaginary numbers were used successfully to describe actual phenomena existing in the real world and these so-called hyper numbers are now indispensable to relativity and quantum mechanics. That raises the question of whether these imaginary units have some type of structural counterpart.

Musès (1972) argues that they do. This general hypothesis is supported by the new superstring theories, which extend the ideas of Riemann, Weyl, and Kaluza-Klein. The hyper numbers, Musès suggests, delineate the various dimensions of the mind. Since $i = \sqrt{-1}$, we could diagram this as a square, with each side having a length of i; the area of the square would be -1. This square would have the same "size" as a square with sides equal to the unit 1 but would exist in inner space.

The creation of the next higher dimension introduces e, which $= \sqrt{1}$, but $e \neq 1$, i, or -1. This alternative "square" is at right angles to the square of the previous dimension, and just as e is "beneath" the integer 1, i is in a plane perpendicular to -1 and one dimension "beneath" it (that is, hyperspace).[8] This hypothetical interdimensional square might be diagrammed as follows:

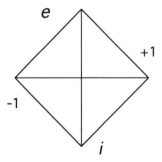

Fig. 11.5. An imaginary square based on Musès (1972), where:
  i *is the imaginary number* $\sqrt{-1}$
  e *is the imaginary number* $\sqrt{1}$, *but is not 1, i, or −1*
  so that $\sqrt{1} = 1$ *and* e, *but* e ≠ *1*
  and e ≠ *-1*

The imaginary numbers *e* and *i* or the $\sqrt{1}$ and $\sqrt{-1}$, respectively, are different dimensions of 1 and -1 because they are at right angles to the normal direction of the real numbers. Their use suggests a link to the non-Euclidean curved space associated with whirling subatomic particles.

The orthorotation of a three-dimensional object (such as a dreidel or the Earth), or the movement of the Earth through space, sets up its own coordinates in primary space, generates change, and creates both the Einsteinian and the Ouspenskian fourth dimension. As we have seen, Ouspensky postulates that this fourth dimension has two aspects, one mental and one physical. The *mental* aspect of the fourth dimension corresponds to imaginary numbers (such as $\sqrt{-1}$), the hypercube, and the inner realm of consciousness. The *physical* aspects of the fourth dimension are related to time as we know it, to the orthorotational/gyroscopic properties of the elementary particles, and to the Earth.

The present moment contains physical and mental components. These two realms travel concurrently through the progression of time.

The orthorotational component, related directly to the spinning Earth, propels this fourth-dimensional unit into the future, thus leaving the dimension of the past behind. Another way of saying this is that the orthorotation of the Earth (as well as the solar system and galaxy) creates the changing dimension we experience as the progression of time. This is due to angular momentum, or movement perpendicular to itself, and the change from a lower dimension to a higher one.

PENROSE AND THE TWISTOR

Penrose's "twistor" concept[9] offers an interesting perspective on the structure of the fundamental fabric of spacetime. In the late 1950s, Penrose "had the idea that the discrete combinational rules obeyed by quantum mechanical angular momentum (e.g., a spin network scheme), could be used as a starting point for building up a spacetime network."[10]

Penrose's theory goes into the world of subatomic particles to describe the basic building blocks of the universe. Influenced by the abstract mathematicians Plücker, Sophus Lie (Lie groups), Cayley, Minkowski, Weyl, Kaluza, and Klein, Penrose suggests that imaginary numbers when multiplied by themselves create the real numbers. Instead of using Einstein's four dimensions to describe "compactified" spacetime, Penrose proposes four complex numbers, each having a real part and an imaginary part, so that in essence eight numbers are needed. "When the imaginary numbers are added to the mathematical loom, they are like the way [hyper]threads are woven at right angles across the real number threads. If the two separate types of numbers are to be properly combined to make complex numbers, they must be interwoven like the warp and woof of threads in a finished cloth."[11] "The actual spacetime we inhabit," Penrose writes, "might be significantly regarded as a secondary structure arising from a deeper twistor-holomorphic reality."[12]

Penrose is trying to create a model that takes into account gravity

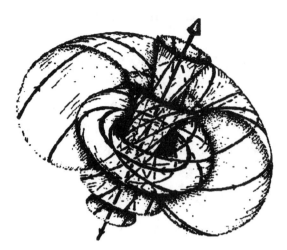

11.6. The twistor

and the nature of space, which he sees as quantized, resulting in eight dimensions.

- three space dimensions
- two angular directions of motions through spacetime
- energy
- spin of the motion
- polarization

According to Penrose, the massless twistors are the fundamental entities that inhabit space. Forward explains in his article on Penrose that space itself is twisted around and shaped by these dimensions: "Mathematically, a twistor is closer to being a square root of a particle."[13] The twistor is extremely small as compared to the size of the tiny electron. Analogous to the graviton or Higgs boson or Maxwell's vortices, the twistor perhaps may be used to portray the fundamental fabric of the ether.

Penrose and his co-workers have been able to demonstrate that combinations of certain types of twistors produce objects that behave like elementary particles. A single twistor can produce one of the known massless particles such as a photon, neutron, or graviton that always travel at the speed of light. Twistor combinations can produce particles like electrons. Three twistors in various combinations theoretically can produce the building blocks of the nuclei of atoms.[14]

Penrose's ideas are appealing, as they make good use of the principle of orthorotation while balancing out the spacetime continuum with imaginary and real numbers. Just as two imaginary numbers when multiplied together create a real number, higher dimensional properties of space create or generate matter in time.

This idea is somewhat compatible to Fredkin's description of "the fundamental level." In Fredkin's model, "[B]oth space and time have a graininess to them" that cannot be broken down further. "At rock bottom . . . people, dogs and trees and oceans . . . are more like mosaics than like paintings. . . . At this level, the universe *is* a cellular automaton, in three dimensions—a crystalline lattice of interacting logic units, each one 'deciding' zillions of times per second whether it will be off or on at the next point in time. The information thus produced . . . is the fabric of reality, the stuff of which matter and energy are made. An electron, in Fredkin's universe, is nothing more than a pattern of information, and an orbiting electron is nothing more than that pattern moving."[15]

Science must become truly objective in admitting the existence of so-called paranormal phenomena, alter its restrictive definition of time, and also come to grips with the fact that the fundamental structure of space is linked to ether theory, imaginary numbers, and corresponding mental and physical hyperspatial counterparts.

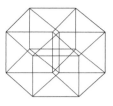

12
PATTERNS OF PROPHECY

The evolution of society is very much like the development of a human being. It builds on past patterns (archetypes). History repeats itself, but with variations.

ALAN VAUGHAN,
PATTERNS OF PROPHECY

An understanding of precognition practically demands a multidimensional structure for time. Clearly, future probabilities (and impossibilities) must exist somewhere. If we introduce the dimension or realm of mind to spacetime, we can see that some of the future exists as goals and hidden trends from the present and past. This can be interpreted in a Freudian way (that is, in a personal unconscious) and as a physical state (that is, the second dimension of time).

Although the future exists one way or the other as probable futures that can be known by some dimension of mind, many probable futures do not manifest themselves on the physical plane. One probable future that did not occur was the assassination of Ronald Reagan, though an attempt was made in March 1981. In the mind of his assailant, Hinckley, the president was going to die. To that extent,

Reagan was already dead, as one of a number of possible futures. Fortunately, this probable future did not become manifest, but it did play out its existence in the inner planes. Former senator Bob Dole will never be president of the United States, although he tried heartily for over a quarter of a century to achieve this end. However, Bob Dole *is* president in some realm of mind as a future probability that simply never made it to the so-called physical plane of reality.

Precognition involves unconscious deductions surfacing into consciousness, appearing as hunches or as dreams. In some cases, some form of foreknowledge is able to prevent a probable event from taking place. For instance, if there is a faulty electrical connection in someone's house and the future fire is dreamt about, if the dream is powerful enough, the person may check the house and possibly locate the problem, thus averting disaster. It is possible that unconscious deductive reasoning based upon psychically received information is the cause of the dream. Thus, a probable future existing in real spacetime (that is, the faulty wiring) along with clairvoyant vision and unconscious thinking would account for the precognitive episode.

With the advances of technology, our relationship to time has changed. For instance, before the invention of the satellite, weather prediction was an uncertain science. Since we now have photographs of moving weather patterns and corresponding evolving computer models, it becomes much easier to predict impending weather conditions. Concerning earthquake predictions, we are also making headway on a number of fronts. We now have computer-linked sensors buried around the globe, as well as infrared satellite photography, which has apparently found a correspondence between idiosyncratic rippled cloud formations and impending earthquakes, and other sophisticated technologies that seismologists continue to develop, resulting in our gaining lead times on earthquakes anywhere from ten to fifteen seconds to possibly hours or days.

With change in perspective comes a change in the knowledge of the future and the structure of time. Thus our relationship to time also changes. The relationship of precognition to synchronicity and the repetitive patterns of human culture become clearer with a discussion and analysis of the truly astounding coincidences surrounding the deaths of Abraham Lincoln and John F. Kennedy.

A B E   L I N C O L N
J. F.   K E N N E D Y

The following is a list of similarities:

1. They both have the same number of letters in their last name.
2. Conspiracies were involved in both cases.
3. Civil rights was an important issue in both administrations.
4. Both men were presidents who were assassinated.
5. Both men were shot in the back of the head.

J O H N   W I L K E S   B O O T H
L E E H   A R V E Y O   S W A L D

6. Both assassins have the same number of letters in their name. Both men are known for three names. Booth shot from a theater and ran to a warehouse. Oswald shot from a warehouse and ran to a theater. This would be an inverted series according to Kammerer's delineations. Both men were killed before a trial took place.
7. Kennedy was shot in a Ford motor vehicle; Lincoln was shot in the Ford Theatre.
8. Premonitions of death preceded each assassination.
9. Kennedy's secretary's name was Lincoln. (This is an inverted series.)
10. Their administrations were one hundred years apart.
11. Both men were followed by a man named Johnson. Andrew Johnson was born in 1809, Lyndon B. Johnson in 1909. (The similarity in the last name would be evidence for a homologous series.)

Numerous articles have been written about these coincidences. Rampa suggests that Kennedy was the reincarnation of Lincoln's spirit, and that the oversoul, of which Lincoln was a manifestation, had not completed its task on Earth. Therefore, it returned to finish what it had begun.[1]

In looking over these coincidences, we must first ask whether these similarities are simply chance events. If they are merely a by-product of the probability laws of large numbers, then the discussion is at an end. Given so many human events, some are bound to be coincidental. However, if meditation upon this series of parallel occurrences reveals more mysterious connections, then some important theories on the structure of spacetime may come to light. First, the relationship of synchronicity to precognition becomes apparent. The events surrounding Lincoln's death served symbolically as a warning and even description of Kennedy's death. The energy surrounding the previous event was resurrected when a similar pattern emerged.

Utilizing his ideas of a collective psyche structured with time-transcendent archetypes, Jung suggests that not only do archetypal patterns "collect events around themselves" during emotionally charged times in history, but they can also "travel across time"[2] so as to reappear with synchronistic precision.

Alan Vaughan applied his idea concerning a "pattern of prophecy" to this presidential coincidence, using a clear-cut actual case to back up the concept.[3]

KENNEDY   JOHNSON   NIXON
LINCOLN   JOHNSON   GRANT

Noticing that both Nixon and Grant had the same number of letters in their names, Vaughan decided to study Grant's administration so as to make predictions about Nixon's. Since scandal had tainted Grant's term in office, Vaughan predicted scandal during Nixon's term. He sent this prediction to the Central Premonitions Registry two years before Watergate, and at a time when Nixon was at the height of his glory. As president, Nixon had just spoken to a man on the Moon, had visited China, and was attempting to end the war in Vietnam. After this prediction was proved to be correct, Vaughan continued his analysis:

1961–63	1963–69	1969–74	1974–77
KENNEDY	JOHNSON	NIXON	GERALD FORD
LINCOLN	JOHNSON	GRANT	RUTHERFORD HAYES
1861–65	1865–69	1869–77	1877–81

Earlier in this treatise, the concept of the "kicker" was introduced: an added synchronicity that serves to emphasize the astonishing accuracy of these kinds of experiences. In reviewing this material in 1981, physics professor E. Gora commented that this was an example of "God playing tricks with us." This group of synchronicities is replete with kickers, for not only do the Ford names line up, but also both these men were the only presidents (to that date) to have been appointed to the presidency! Ford, like Hays, was appointed, but for Ford, it was to the vice presidency when Spiro Agnew resigned, and for Hayes, it was to the presidency. In essence, though, Congress knew that Agnew's replacement would become president, as Nixon's days in that capacity were numbered.

These coincidences have also been linked to the twenty-year death cycle of the presidents. The following men died in office; their elections having occurred every twenty years:

Lincoln	1860	Assassinated
Garfield	1880	Assassinated
McKinley	1900	Assassinated
Harding	1920	Illness
F.D. Roosevelt	1940	Illness
Kennedy	1960	Assassinated

This twenty-year cycle was probably used by Jeane Dixon—in 1956—to predict the death of the president who would win the 1960 nomination.[4]

Inherent in this discussion are principles concerning synchronicity, precognition, the structure of the collective psyche, and the structure of time. As stated earlier, precognitive episodes are involved with future probabilities, not certainties. However, if there really is a synchronistic link between the deaths of Lincoln and Kennedy, then we must take the next bold step and concede the possibility that predestined forces order seemingly random human events. Perhaps archetypal psychic patterns (involved with assassinations) existing in the psyche of the species *Homo sapiens* "collected the two events around themselves," as they "traveled across time" to reappear with astounding recurring precision.[5] Thus, Jung's so-called acausal mechanism for synchronicity can be seen in a new light. The two events are not connected in a normal causal way, but are linked by a psychic pattern that exists beyond time and space. This realm is a mental one not subject to "normal" spacetime constraints.

GURDJIEFF'S LAW OF THE WILL

With any discussion of precognition, it is important to reiterate that we really do have free will. However, we rarely utilize it. Gurdjieff writes that the world is run by the "law of the will," which has three aspects:

1. God's will, which rules life and death, disease, and natural disasters.
2. Big Brother's will (term borrowed from George Orwell, which of course Gurdjieff did not use), that is, the will of people and entities more powerful than you, such as the government, big business, and your boss.
3. Your own will, which according to Gurdjieff is rarely exercised. A person's genetic and astrological makeup sets up a general

trend for that person, but individual decisions and ultimate free will are always there as a potentiality.

Certainly we cannot be separated from the forces of the solar system. Every time we look at a watch to see what time it is, we are correlating ourselves to the position of the Earth with regard to the Sun. Another way of asking what time it is to inquire: "Where is the Earth in relationship to its rotation with respect to the Sun?"

For all intents and purposes, there is nothing whatever uncertain about the future with respect to the movement of the planets and stars. In fact, much of the future has already been written, literally *millions* of years' worth! We just haven't seen it yet, as it takes such an enormous amount of time for light from stars in our own galaxy to reach us, let alone light from distant and even nearby galaxies. For instance, light from the far reaches of our own Milky Way can take 100,000 years to reach us, and the closest galaxy to us, Andromeda, is two million light-years away!

The inertia of the stars and galaxies throughout every region of space has a timeless quality, because even the present is highly influenced by occurrences that happened hundreds of thousands and millions of years ago. When we look out at the stars, we are looking at the history of the past, as it takes "time" for light to travel from the stars to the present position of the Earth. Ironically, we really have no way of knowing the present state of things in the cosmos because of this great time delay. Most of our theories on the structure of the universe are based on information that is millions of years old. That's a long time ago, and what that means is that hard-nosed scientists simply have faith that what happened way back when is similar to what's happening right now. But really, it's just a guess, based on—my heavens—faith!

Since it is clearly established that the vast bulk of our earthly environment is essentially predetermined (such as the seasons, eclipses, and visits by comets), then it may also be possible that (some) human events are as well. Some predetermined truths are listed below:

1. The human zygote grows into an embryo and eventually into a fully developed child.
2. The child grows and changes shape. As a person matures, his or her hair turns gray and his or her whole system winds down. Eventually the person dies.
3. The seasons return every year.
4. Many people will watch television tomorrow.
5. Some birds migrate annually.
6. Humans will build space colonies on the Moon.

The future in terms of human events is not carved in stone, but many future events have a tremendous amount of momentum behind them. In the case of predicting John Kennedy's assassination, it is clear that at least one person (Lee Harvey Oswald) knew well in advance that Kennedy could be or would be assassinated (assuming Oswald was the culprit). Thus, that mental energy existed as a strong probable future before it became manifested on the Earth plane.

Things get more complex if we relate predetermination to such events as the sinking of the *Titanic* and the fact that this case was "precognized" fourteen years before the ship was even built. Given that predetermination is a powerful force in our environment, can we say that the demise of the *Titanic* was "preordained"? This book does not argue that at all. Adolf Hitler could have won the war had he not invaded Russia. Al Gore would have been president if the butterfly ballots had not been printed. A nuclear accident could have happened in Cuba, triggering a holocaust there when John Kennedy was president. However, there still are predisposing forces.

In discussing the problem concerned with synchronicity and precognition, Carl Jung turns to the concept of teleology. The teleological principle holds that the end is implied in the beginning. All goal-directed behavior involves this principle. Implied in the acorn is the future oak tree, which to some extent directs the growth of the acorn. We can see that all of biological life is teleologically determined. The phenomena of growth, adaptation, evolution, and reproduction are all witness to this fact.

The point concerning human events is to consider where "cosmic" predetermination and our own willpower leave off and randomness, chance, and uncertainty enter in. The future is to be looked at as a series of probable futures. There was a probability that Kennedy would be assassinated, just as there was a probability that Tony Blair would be. In Kennedy's case, the probability became an actuality. Ouspensky would argue that the future is determined by the fifth and sixth dimensions of time, that is, when the "could have beens" and "the all possibilities existing in eternity" manifest themselves on the physical plane. This deeper dimension of time certainly exists in our minds. Our plans for the future exist as imagination. Yet these cognitive structures should also, in fact, exist in some physical sense, in a realm we are now calling hyperspace.

Evidence for telepathy, synchronicity, and precognition strongly suggests the presence of the universal mind, one that all humans have access to. According to this theory, at the periphery we are each individuals, but at deeper levels, all psyches transcend individuality and blend into the collective mind. In this model, telepathy would simply be the ability to tap into what is already present at deeper layers. What one person thinks occurs as a faint vibration in the collective psyche of all people.

As we have seen, Jung suggests that this collective psyche is organized by means of archetypes and that, to some extent, this overarching mental structure influences everyday events. Progoff relates the archetypes to deep sources in nature:

> The archetypes are a next step in the unfoldment of the image that is attached to the instinct just before it reaches the point where its characteristics become distinct. . . . Within the terms of the evolution of the animal kingdom, the basic psychological images arise from the patterns of behavior that are native to the human species,

drawing both their symbolic tendencies and the great motor power of their energy from primal sources deep in nature.[6]

Progoff goes on to suggest that we as microcosms reflect the workings of the macrocosm. If we give credence to the occult law "As above, so below," the assumption could be made that if we think, so does the cosmos. Evidence for pure precognition suggests the presence of a mental component existing in hyperspace and the second and third dimensions of time. Certain human events may be conceived days, months, years, or even generations before they actually occur on the physical plane, but these events exist in this realm as a series of probabilities, not certainties. Martin Luther King's speech in which he envisions real equality and respect for all people is a case in point. We all hope for a day when true equality between the sexes, peoples, and nations will arise. Our goal is to shape a worthy future and oppose negative and detrimental trends.

ASTROLOGY AND PLANETARY FORCES

The wise man rules his stars. The fool is ruled by them.
                                    MAX HEINDEL, 1909

History is filled with famous astrologers. This list includes just about every founding father of modern science: Ptolemy, Copernicus, Pythagoras, Galileo, and Newton. Kepler, in fact, was the official court astrologer to Emperor Rudolf in Prague. In modern times, Joan Quigley was court astrologer to President Ronald Reagan and his wife, Nancy. The fact that astrologers abound in the present day suggests that many people feel that life on the Earth is not separate from the rest of the solar system. That is really the key point regarding the underlying precepts of astrology. In fact, if we take out a telescope and simply look at the night sky, it becomes quite obvious that some of the great bodies of the solar system are very close by, for there in stunning glory, one can easily see the craters on the Moon, Jupiter with four of its moons, or the spectacular rings of Saturn.

Astrology is a synchronistic practice and an empirical discipline that links the position of heavenly bodies to either different personality types or the occurrence of global events, such as war, prosperity, and famine. Knowledge of the future positions of the planets becomes the basis for a fairly straightforward procedure of prediction. Expanding these thoughts, we can see that astrological doctrine hypothesizes a relationship among the individual human mind, the collective psyche, and the structure of the solar system. If there is tension in the stars, it is felt here on Earth.

The astrologer Nostradamus (1503–1566) is considered by many to be the world's greatest seer. Perhaps he could be called a scientist of prophecy. Concerning the coming of Napoleon, he wrote that "an Emperor . . . rase-tete [of short cropped hair] . . . will be born near Italy [Corsica] whose empire will cost France very dear."[7] Another prediction of his has been linked to the rise of Adolf Hitler, whom Nostradamus apparently called "Hister." Certainly, most of his predictions are ambiguous at best and written in symbolic form. Nevertheless, it is quite pos-

Fig. 12.1. Saturn photographed from the satellite Cassini

sible that Nostradamus was able to link clairvoyant vision with future planetary configurations based upon astrological doctrine.

> Some stars scintillate with a strange glow . . . like the eyes of fantastic animals in the dark jungle of the sky. . . . This star worship . . . is essentially derived from the quality of the *light* of the star. . . . With the Sun and the Moon, and probably later with the brilliant stars, man also feels a vague identity. He feels them, tries to become ever more one with them, to become instinct with their essence.[8]

At first the ability to prophesize an event that may take place hundreds of years in the future appears incomprehensible. However, as we have seen, a clearer understanding concerning the structure of time helps to make precognitive or prophetic revelations more plausible. The structure of time is directly related to, and based upon, the movement of the Earth through space. Therefore, the past is correlated to where the Earth was; the future, to where it will be.

One thing that is immediately apparent is that the movement of the Earth through space (for all practical purposes) is completely

Fig. 12.2. Faust 1652, Rembrandt

predetermined. In fact, the movement of the entire solar system has maintained a stability and predictability for thousands if not hundreds of thousands or even millions of years. It is also likely that the momentum of the planets and the sun will continue unhampered upon their path for eons to come.

Still, there is a certain amount of unpredictability in the movement of the stars, and asteroids and meteors do occasionally strike the Earth, often with catastrophic results. It is believed, for example, that the great crater in Arizona may have caused the extinction of the dinosaurs sixty-five million years ago. Much more recently, the explosion in Tunguska, Siberia, which in June 1908, took out an area the size of Rhode Island may have been caused by an asteroid skimming the Earth. But overall, at least in terms of the short spans of a few hundred or a few thousand years that generally relate to human events, the Earth and the solar system will most likely maintain a stalwart stability.

Extrapolating from Dane Rudhyar's book *The Sun Is Also a Star,* we come up with the following traditional symbolic scheme for the planets:

Sun: Conscious
Moon: Unconscious
Mercury: Fastest planet—communication and sex
Venus: Female
Mars: Male
Jupiter: Largest planet—expansion, the Mother
Saturn: Last of inner planets—restriction, the Father

Then come the outer planets, which represent higher harmonics of the last three inner planets: The inner planets are the ones that the ancients knew about because they were noticeable to the naked eye: the Sun, Mercury, Venus, the Moon, Mars, Jupiter, and Saturn. The Sun and the Moon are considered "planets" for simplicity sake in this discussion. The last three are Mars, Jupiter, and Saturn. Then there are the outer planets, in order, Uranus, Neptune and Pluto. These represent "higher harmonics" of the last three inner planets, Uranus is the higher harmonic of Saturn, Neptune of Jupiter, Pluto of Mars. They have, astrologically, the same essential "energy." Since Mars has a bellicose side, so does Pluto. Jupiter is an expansive planet, Neptune is a dreamy one. Saturn has a hard scientific side, so does Uranus. But Saturn is the last of the planets known to the ancients. So what lies beyond Saturn represents a major qualitative change. Uranus stands for rapid change.

Uranus (higher harmonic of Saturn)—rapid change
Neptune (higher harmonic of Jupiter)—new age
Pluto (higher harmonic of Mars)—war[9]

Jung's scientific studies of astrological principles, as well as M. Gauquelin's research, have shown an empirical relationship between the major planets and particular professions or personality types. Based upon a survey of 25,000 professional people, Gauquelin (1967) writes in his book *Cosmic Clocks* that the following planets were at the mid-heaven (highest point in the sky) at the births of people who ended up in the following vocations:

Mars: Scientists, athletes, military men
Jupiter: Military men and politicians
Saturn: Scientists, doctors
Moon: Writers

Gauquelin also noticed that there was a very *low* incidence of other pairs:

Painters, musicians, and writers did not correlate with Mars
Doctors did not correlate with Jupiter

Painters and writers did not correlate with Saturn

Athletes did not correlate with the Moon[10]

The strongest correlations, above a million-to-one odds, were for military men with Mars or Jupiter and for athletes with Mars. Note that this is an empirical finding; it most likely reflects similar studies that were done by the ancient stargazers who first developed the astrological precepts. It is important, even essential, for critics of astrology to realize that what Gauquelin has done is *test* the premise of astrology, and he found a valid connection. I would urge such critics to replicate these studies or at least carefully read his book before they continue to dismiss these empirical findings. The reason for this is obvious. If there is something to astrology, if Newton, Copernicus, and Kepler, among others, are right on this point, the implications are indeed profound. Each of us is a reflection of the cosmos, and more than that, the cosmos has a personality of sorts, harmonic and antagonistic energies and curious synergestic, dynamic, and idiosyncratic effects.

To understand the physics of astrology, we need to know a few things. All of the planets lie in an eighteen-degree band around the Sun, which is called the ecliptic. Picture the Sun set up much like the planet Saturn with its rings. All of the Sun's planets are in the same basic plane, but in different orbits extending out to Pluto. If that band is continued to the Milky Way, that is where the zodiac is. It is composed of twelve particular constellations encircling the Sun in the same band as the ecliptic. These constellations, in order, are as follows: Aries, Taurus, Gemini, Cancer, Leo, Virgo, Libra, Scorpio, Sagittarius, Capricorn, Aquarius, Pisces. Naturally, there are many other constellations, such as Orion, Cassiopeia, Vela, Ursa Major, and so on, but these do not lie in the ecliptic. Each sign of the zodiac is associated with a particular planet, although in three cases, Aquarius, Scorpio and Pisces, these signs have co-rulers, that is, two planets ruling them. The idea essentially is that the twelve sectors of the zodiac are an extension of twelve rays of energy emanating from the center, which for us is the Earth. The chart below depicts the various planets that rule the twelve constellations that lie in this ecliptic. The implication, of course, is that the energy of the Sun, sunlike qualities, match the energy of the constellation of Leo, which truly looks like a lion. Scorpio is a constellation that resembles a scorpion. This is an aggressive sign, and it therefore is tied to Mars, the god of war. Taurus in the sky resembles the steady bull, and this constellation is therefore aligned with the solid and warm planet Venus, and so on. That is how the planets are linked to the constellations.

This is a very complex typology that lies at the basis of numerous theories of personality including Carl Jung's four types, Thinking, Sensing, Intuitive, and Emotional (Fire, Earth, Air, and Water); and Raymond Cattell and Gordon Allport's idea of cardinal and central traits, which stems from the concept of ruling planets in a chart (for example, the person is mercurial, saturnine, a lunatic, etc.)—and powerful signs dictated by the number of planets in the sign and/or the position of the sign in the chart, such as on the mid-heaven (at the top of the chart) or as the rising sign, at nine o'clock corresponding to the horizon at the time of birth (the age of Aquarius, stubborn like a bull [Taurus], for instance).

Astrological readings are laid out upon a chart that is set up like a clock with twelve houses corresponding to the twelve constellations of the zodiac. The location of the Sun at the time of birth determines a person's Sun sign. A person is an Aries or Capricorn or Virgo because the Sun is in this particular constellation when he or she was born. The location of the Moon and the planets at the time of birth are also noted on the chart. One of the key goals in understanding a horoscope is to locate the most powerful planet in the chart. This is called the *final dispositor.* A dispositor is simply the planet or its essence that rules a particular sign. The signs and their ruling planets are shown in the list below.

PLANET RULES	SIGN(S)
☉ Sun	♌ Leo
☽ Moon	♋ Cancer
☿ Mercury	♊ Gemini and ♍ Virgo
♀ Venus	♎ Libra and ♉ Taurus
E Earth	♉ Taurus
♂ Mars	♈ Aries and ♏ Scorpio
♄ Saturn	♑ Capricorn and ♒ Aquarius
♃ Jupiter	♐ Sagittarius and ♓ Pisces
♅ Uranus	♒ Aquarius
♆ Neptune	♓ Pisces
♇ Pluto	♏ Scorpio

Fig. 12.3. The signs of the zodiac and their ruling planets

In astrology, for convenience' sake, both the Sun and the Moon are referred to as planets. Sometimes finding the most powerful planet is easy; more often it is complex. Astrology weights the planets differently by separating them into three groups. The Sun and Moon are in the first group; the inner planets Mercury, Venus, Mars, Jupiter, and Saturn are in the second group; and the outer planets, Uranus, Neptune, and Pluto, are in the last group. The Sun rules Leo and the Moon rules Cancer. All of the inner planets rule two signs each, and the outer planets co-rule three signs:

- Saturn and Uranus co-rule Aquarius
- Jupiter and Neptune co-rule Pisces
- Mars and Pluto co-rule Scorpio

In an astrological sense, the co-ruling planet for each of these three signs is seen as a higher harmonic (more advanced vibe) of the first planet: Uranus is the higher harmonic of Saturn, Neptune the higher harmonic of Jupiter, and Pluto the higher harmonic of Mars.

The planets are seen almost in a musical sense. Thus the first seven,

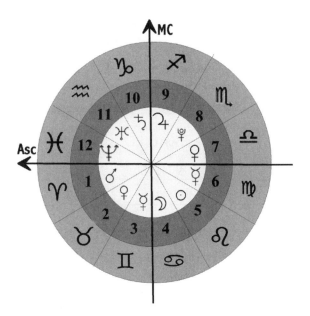

Fig. 12.4. The zodiac, with the ruling planet for each sign

the Sun through Saturn, correspond to the first chord; and the next three, the outer planets, Uranus, Neptune, and Pluto, correspond to the same chord at a higher range. The outer planets are rhythmically linked to the last three inner planets.

Each sector of the zodiac is considered to generate a particular vibration, represented as its theme. For instance, the first house, right below the horizon, rules personality; the second, financial affairs; the third, siblings; the fourth, the home; and so on. If a planet is in its own sign, its energy is greatly magnified, if the planet is in a friendly sign ruled by a compatible planet, it becomes "exalted." If the planet is in a somewhat unfriendly sign ruled by a somewhat incompatible planet, it is in its "detriment"; and if the planet is a sign with which it is incompatible, that planet is in its "fall."

For instance, the Moon rules Cancer. Any planet that is in Cancer at the time of birth will be modified so as to emphasize its imaginative/lunar aspect. Jupiter, which is an expansive planet, is exalted in Cancer, but Mars does not like it there, and thus Mars is in its fall if it is in the sign of Cancer. Mars rules Aries, so any planet in Aries will display its more masculine side. The Sun is exalted in Aries, but Venus, a feminine planet, is in its detriment there, and Saturn, which is in constant conflict with Mars, is in its fall.

SIGN	RULER	EXALTED	DETRIMENT	FALL
Aries	Mars	Sun	Venus	Saturn
Taurus	Venus	Moon	Mars	
Gemini	Mercury	Jupiter		
Cancer	Moon	Jupiter	Saturn	Mars
Leo	Sun	Saturn		
Virgo	Mercury	Jupiter		
Libra	Venus	Saturn	Mars	Sun
Scorpio	Mars and Pluto	Venus	Moon	
Sagittarius	Jupiter	Mercury		
Capricorn	Saturn	Mars	Moon	Jupiter
Aquarius	Saturn and Uranus	Sun		
Pisces	Jupiter and Neptune	Venus	Mercury[11]	

One could ask why Saturn is in its fall in Aries, which is ruled by Mars, yet Mars is exalted in Capricorn, which is ruled by Saturn. Inherent in a question like this is the issue of how much of astrology is arbitrary. Most likely some of it is, but overall there is an inherent logic to the scheme. Again, please see such books as *How to Learn Astrology* by Marc Edmund Jones for a more in-depth explanation. Concerning the question above, the answer may have to do with the mythological relationship between Mars and Saturn. If Saturn represents the father, the son, Mars, may be exalted in the father's sign; the father may indeed be in his fall in the son's home.

The next things that are taken into account are the "aspects," or the geometric relationships among the planets. There are two types of easing relationships: 1) conjunctions, whereby two or more planets are close together, usually within ten degrees, and 2) the trine, which is a sixty-degree angle. There are also two types of difficult relationships: 1) squares, which are ninety-degree angles, and 2) oppositions, which are planets that are 180 degrees apart. All of these aspects are within about a ten-degree tolerance. If Venus is opposed to Mars, there may be a problem with sexual identity. If Jupiter is opposed to Saturn, this suggests trouble between the parents. If Saturn is square Mercury, a person may have difficulty restricting his or her finances.

APPLYING THE PRINCIPLES

The practice of astrology will become clearer if we look at a few examples. Even at the height of his impeachment proceedings, President Bill Clinton still had a 65 percent approval rating. The fact that he was able to effectively rule the country at that time was nothing short of amazing. The man has enormous talent. This ability could be explained in part by his chart, shown below. The Sun is in its own sign, Leo, making it a very powerful planet in his horoscope. He also has Venus in Libra, which is in its own sign as well. Therefore, these are the two most powerful planets in his horoscope. The Sun directly rules Pluto, Mercury, and Saturn, because these three planets are in Leo. Indirectly, the Sun also rules Uranus because that planet is in Gemini, a sign ruled by another planet, Mercury, that is already ruled by the Sun. So the Sun is Clinton's first final dispositor. His sign of Libra is ruled by his other final dispositor, Venus. This planet will rule all of the planets that are in Libra: Mars, Neptune, Jupiter, and itself. Venus will also rule the Moon because the Moon is in the other sign that Venus rules, Taurus.

Libra was on the horizon when Clinton was born. That's what is known as his rising sign. Rising signs are powerful by definition;

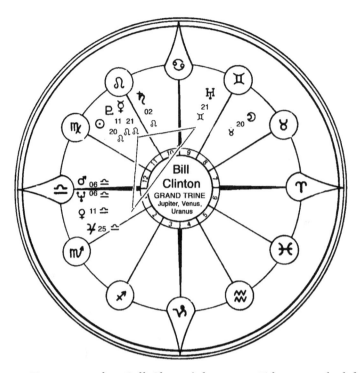

Fig. 12.5. Former president Bill Clinton's horoscope. Libra is on the left as the rising sign. It contains Mars, Neptune, Venus, and Jupiter. At eleven o'clock is Leo, which also contains four planets, the Sun, Pluto, Mercury, and Saturn. Uranus, at one o'clock, sits in Gemini, and the Moon, at two o'clock, is in Taurus. The Sun is trine Uranus and also trine Jupiter. Many planets are favorably placed.

however, in his case, he also has four planets in Libra, one of them being the ruling planet of Libra. Therefore, Bill Clinton has two very powerful signs in his chart. Even though he was born a Leo, he is just as much a Libra, and a very powerful Libra at that. The Sun trines one of the planets in Libra, Jupiter. This magnifies even more so the positive aspects of these two essentially benevolent signs.

Note also that Mercury and Saturn are both in Leo and that the Sun trines Uranus in Gemini. In the first instance, Saturn, the planet of discipline, is linked to the planet of finances, Mercury. Uranus, the planet of rapid change, in the highly fluid sign of Gemini, suggests unseen potential catastrophes—in his case, impeachment. This difficult position, however, is offset, because Uranus in Gemini is trine the benevolent Sun, and the Sun rules Uranus through its influence over Mercury, the ruler of Gemini. Because his Sun is so well placed, it suggests triumph in the end.

Mars in Libra is associated with Clinton's tendency to be highly interested in the opposite sex. The fact that Libra is ruled by Venus indicates that Clinton could be dominated by women, but at the same time not be threatened by them because he himself understands the feminine component. First off, this lion married a strong female in Hillary, who is now a senator and recently a presidential candidate. The intern he had the affair with clearly held his attention, too much so, considering her age and underling position. On the other hand, Clinton appointed a woman to the Supreme Court and made Madeleine Albright Secretary of State, the first female ever to hold that position.

Sigmund Freud, too, has a most unusual chart, one that explains his obsession with the Oedipal complex. In Freud's chart, Mars, the god of male sexuality, sitting at the bottom, is opposed by every other planet. Note that Mars is intercepted in Libra. Certain constellations, or in this case signs of the zodiac, are not always visible at the time of birth due to the tilt of the Earth. These are called intercepted signs. If there are planets in those signs, the energy from those planets is somewhat trapped. This is what is happening in Freud's case. Essentially, every planet is opposite to Mars. This magnifies greatly the energy of Mars, but at the same time its energy is trapped with no easy outlet. Note that Venus, his feminine side, is also intercepted at the top of the chart, in Aires. The sexual affection Freud had for his mother had to be repressed in the male sign of Aries, and at the same time the jealous feelings for his father, who was bedding his mother, blocked in the feminine sign of Libra, also had to be thwarted. This configuration can easily be seen as symbolic of Freud's Oedipal complex.

We also see several other things at play here. With such powerful planets and signs intercepted, this in and of itself suggests that

powerful unconscious forces at play. Simultaneously, we see the reciprocal relationship with Mars ruling Venus through Aries and Venus ruling Mars through Libra. These are powerful male and female sexual
energies being interchanged in a horoscope with no easy outlet.

In terms of his Sun sign, Freud is a Taurus. This is a staid sign
that gives the doctor a certain stability that helps him override a rather
tumultuous, emotionally charged chart. His final dispositer is Mercury,
which is in its own sign of Gemini, a fluid sign (mecurial) that also
symbolizes duality, male and female, conscious and unconscious. As the
fastest planet, Mercury is also the planet of sex and communication,
which fits very well with Freud's life and work.

The three most powerful signs in his chart are Taurus, his Sun
sign; Gemini, where his most powerful planet lies; Mercury, and Libra,

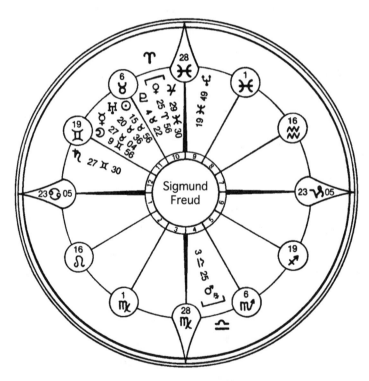

*Fig.12.6. Sigmund Freud's chart. Notice how Mars, which symbolizes
masculinity and male sexuality, is on one side of the chart and all of the other
planets are on the other side.*

which houses Mars alone in it against, essentially, all the other planets. Freud, of course, is the most important explorer in the realm of the unconscious. The fact that Libra, a feminine sign, under the horizon, at the bottom of the chart (the realm of the unconscious), is intercepted, matches well with Freud being the father of psychoanalysis.

In the case of Marilyn Monroe, (chart not shown), she's a Gemini, which means that her Sun is in Gemini, but she also has Mercury in Gemini, which would make Mercury her final dispositor, because Mercury rules Gemini. Thus, she is a "double Gemini." Marilyn also has Venus, the sign of her femininity, in Aries, which is ruled by Mars (men). This combination would suggest a highly mercurial nature displaying many different personalities or sides to her personality and multiple sexual liaisons, with a life ultimately dominated by men.

Other complex cases, that also have intercepted (blocked) signs, include the charts of Einstein and Hitler. In Einstein's case, his Moon is intercepted in Sagittarius. First off, the Moon, although technically exalted in Sagittarius, is quite happy there because the imaginative nature of this planet is allowed to expand in that sign. But because Sagittarius is intercepted, this would suggest a strong tendency to mull thoughts over for long periods of time before the associated ideas finally find their way to an outlet.

Einstein is a Pisces because his Sun is located there. Above the sign are the two co-rulers, Jupiter and Neptune. So, the next step is to find where those two planets are located. Let's follow Jupiter first.

The first thing we notice is that Einstein has three planets in Aries: Mercury, Saturn, and Venus. We also see that Einstein is a Pisces, because that is where his Sun is. The combination of these two signs suggests an odd combination of being that of a dreamer (Pisces), yet a leader at the same time (three planets in Aries). I think we can see how that fits. However, the Aries nature is tempered by the qualities within that sign, that is, the three planets: Mercury, Saturn, and Venus (writing ability, interest in science and love—he was married twice— his interest in violin might also be linked to these three planets in

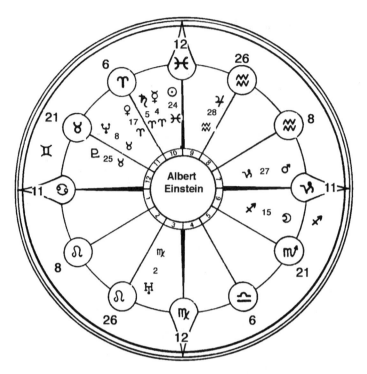

Fig. 12.7. Einstein's chart

Neptune				Uranus		
Jupiter	Mars	Saturn	Mercury	Saturn	Venus	Jupiter
Pisces	**Aries**	**Capricorn**	**Virgo**	**Aquarius**	**Taurus**	**Sagittarius**
Sun	Saturn	Mars	Uranus	Jupiter	Neptune	(Intercepted)
	Mercury			Pluto		
	Venus					Moon

The planets above the zodiac signs in the chart above rule those signs by definition. Planets below the zodiac signs are in those signs appearing in Einstein's chart.

Aries, Venus softening the Saturnian/stern side). Yet at the same time, at three o'clock on the chart we see that his powerful Martian character does not have full reign because his Mars is in Capricorn, which is ruled by Saturn. Thus, Saturn is one of his final dispositors. The strict Saturnian nature tempers the Martian qualities of Aries, which rules over three planets, one of which also happens to be Saturn, and so, we see a reciprocal relationship. Mars is in Saturn's sign of Capricorn and Saturn is in Mars's sign of Aries. Saturn has the edge because it rules Mars, the ruler of Aries and is also in Aries.

Turning to Jupiter, the main ruler of his Sun sign Pisces (at one o'clock), we see that it is in Aquarius. This configuration correlates to Einstein's expansive (Jupiter) interest in the New Physics (Age of Aquarius). However, Aquarius is ruled by two planets, Uranus and Saturn, which are eventually ruled by Mars and Saturn in a reciprocal relationship (Uranus in Virgo is ruled by Mercury, which is ruled by Mars in Aries, which is ruled by Saturn). Ultimately, the planets Mars and Saturn are his final dispositors. Therefore, the expansive energy of Jupiter (the major planet of his Sun sign) is ultimately tempered by a Martian/Saturnian influence. Symbolically, the combination of Mars and Saturn is usually a dark relationship: Einstein is the father, unfortunately, of the A-bomb.

Now let's turn to the other co-ruler of his Pisces Sun sign, Neptune. It sits in Taurus, along with Pluto, so it is ruled by Venus, but Venus is co-ruled by Mars and Saturn. Einstein's Sun is then ultimately co-ruled by Saturn and Mars, tempered by the Piscean character. The natural amiable and imaginative nature of the Pisces is expanded in a peculiar way because his Moon, which is intercepted in Sagittarius, is overridden by the Martian leadership qualities of Aries and the disciplined scientific bent of Saturn. Through strong powers of imagination (Moon in Sagittarius, Jupiter in Aquarius), Einstein uncovered rather strict laws of the universe (Saturn as a final dispositor), but he also had a flexible mind (Pisces character). He stubbornly opposed the move by quantum physicists to make the ultimate substrate of the universe ambiguous (Heisenberg's principle of uncertainty), as Einstein felt strongly that "God doesn't play dice," that is, that the universe would not be ultimately arbitrary. That bent is Saturnian.

The theories of the kindly Einstein led humanity to the ultimate weapon, the atom bomb. This can be seen from a mystical point of view, in a number of ways: the reciprocal relationship of Mars and Saturn, which together can generate a very negative force, and the way the Piscean character and staid quality of Taurus are offset by housing both Pluto and Neptune, planets that have a dark scary side when combined in the same sign.

In Hitler's horoscope, the Sun sign is right on the cusp between Aries and Taurus, at 0°47 in Taurus. Technically, he is a Taurus, a feminine yet stubborn sign, but he is also strongly pulled by the masculine sign Aries because the Sun was extremely close to Aries when he was born. Since his Sun sign Taurus is intercepted, the powerful staid and feminine-based energy is locked up, while the masculine side (the Aries nature) has easier access to expression. All this suggests a deep inner frustration because his Sun sign is intercepted, as well as an ambiguous aspect because the pull of masculine Aries fights the more feminine Taurus.

Mars and Venus are conjunct (next to each other) in the intercepted sign of Taurus. At the same time, Venus rules Taurus. Thus we see a great and complex battle between Mars and Venus. Since his Sun sign is almost Aries, that gives his Martian side a powerful outlet. At the same time, Mars is thwarted by its ruler Venus, which, however, is intercepted in its own sign. All of this is square Saturn, which leads to a defensiveness and difficulty in maintaining close relationships. There is a strong bisexual component inherent in this conjunction.

Saturn in Leo indicates a somber disposition, but also power, particularly with Saturn at the highest point in the chart, at the midheaven, at twelve o'clock. His rising sign, like Clinton's, is Libra, but in Hitler's case the only planet there is Uranus, a sign of revolution. Pluto, the god of war, is in the eighth house, which is the house of death. Hitler seeks to exterminate all non-Aryan races.

Besides the difficult Mars/Venus conjunction, Hitler also has two other unusual conjunctions, the Moon with Jupiter in Capricorn, which would be associated with his megalomania, letting his imagination take over in an expansive capacity, and Neptune, a spacey planet, with the dark and bellicose Pluto, both lying in the mercurial sign of

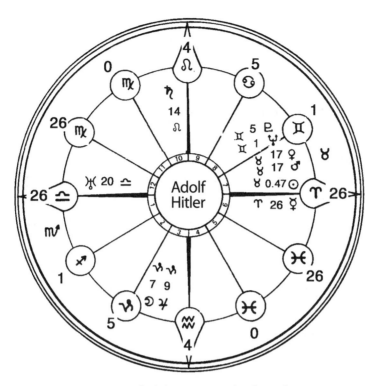

Fig. 12.8. Hitler's horoscope and ruling planets

Venus	Mercury	Sun	Venus	Saturn	Mars
Taurus	Gemini	Leo	Libra	Capricorn	Aries
[Intercepted]	Neptune	Saturn	Uranus	Moon	Mercury
Sun (barely)	Pluto			Jupiter	
Venus					
Mars					

ASPECTS

Oppositions: ♓ ∞ ☿

Squares: ♄ □ ♀, ♂

Trines: ☉ △ ☽

Conjunctions: ♀ ☌ ♂; ♆ ☌ ♀

☽ ☌ ♃; ☉ ☌ ☿

This is a very complex and unusual chart. Note, for instance, that Hitler has three signs intercepted in his Sun sign, Taurus, yet his Sun is barely in Taurus. He is almost an Aries. A lot of anger and frustration is evident in this chart.

Gemini. With Mercury in Aries, this would be associated with swift military action, while the Pluto/Neptune conjunction and confusion of his masculine and feminine sides would lead to duplicity.

This complex process may seem confusing at first reading. If you are serious about learning this discipline, it is strongly suggested that you reread these passages slowly. The key is to understand the underlying nature of each planet and the meaning of the final dispositer. Of course, this short lesson only taps the surface of astrology. There is much more. Someone like Nostradamus, who spent his life studying this highly complex discipline, could create a chart for a country or even an epoch; he could see what types of energies were at play in the present and what they would be in the future.

By looking at great tensions between major planets during key times in history, such as in the late 1700s, when there were revolutions in America and France, and in the 1860s with the Civil War, or during times of financial turbulence, one begins to understand the predictive aspect to astrology. For instance, the horoscope for Black Thursday, October 24, 1929, when the stock market crashed, shows a grand cross where Saturn opposes Jupiter and both are square Neptune. This is a chart of great stress. Astrologer Robert Gover notices that Saturn is also *afflicted,* in that it is located in a sign not well suited for it, which in this instance is Sagittarius. During another stock market crash, in 1987, Gover notes that Saturn is conjunct Uranus and again is afflicted in Sagittarius.

In the early 1970s, while I was working in New York City, I had the privilege of taking courses with Zoltan Mason, a master astrologer who had a shop on Lexington Avenue around 60th Street. This was a time when bookstores displayed credible books on this topic. Today, it is difficult to find a serious work on astrology in the mainstream market. Most of the rubbish sitting on the bookshelves is essentially useless in understanding what astrology is really about. If you would like to explore this complex but fascinating field, books by such authors as Marc Edmund Jones, George Llewellyn, and Dane Rudhyar are recommended, as they explain its logic in detail.

PLANETS, PERSONALITY, AND
BRAIN STRUCTURE

Each sign of the zodiac falls into one of four types: fire, earth, air, and water. Applied to personalities, these types have been renamed by Jung as thinking, sensing, intuitive, and emotional. Referring to the neurophysiological data on the two hemispheres of the brain, we know that the "thinking" and "sensing" types are more left-brain oriented, where language (thinking), seeing is believing (sensing), and sequential thinking are located. The "intuitive" and "emotional" types are more right-brained, where holistic thinking, picture and face recognition, and creativity are located. I thus make the leap and suggest that the left hemisphere of the brain may be more under the influence of the conscious/predictable Sun and the right hemisphere more ruled by the subconscious/flighty Moon. This is at least true symbolically, but I am also hypothesizing that the various geomagnetic forces of the Sun, Moon, and planets affect the brain's superstructure as well, at least in terms of personality.

It is well known that the Sun, Moon, and, in fact, all of the planets have an effect on the tides. Women's menstrual cycles often correlate to the full moon and the eleven-year solar cycle of the Sun reversing its poles has been correlated with the onset of diseases and changes in the growth of tree rings on the Earth. In 1974, for my master's thesis, I calculated the gravitational attraction of the Moon, Sun, and Jupiter on the Earth, using Newton's formula $m_1 m_2/d^2$ (where m_1 = the mass of the first planet; m_2, the mass of the second planet; and d = the distance between them). If the Moon has a gravitational attraction on the Earth's tides equal to the integer 1, then the Sun has a gravitational attraction of about ½, and Jupiter's force on the tides is approximately 0.000001 percent of that of the Moon. This seems like an insignificant amount until it is realized that such drugs as curare and LSD cause profound changes in the human body and brain when taken at microgram dosages, that is, at millionths of a gram, essentially the same percentage as Jupiter's influence on the tides.

Since the brain is 85 percent water, I hypothesized that just as the planets influence the tides, they could influence the neuronal network in humans as well, the brain acting, in this instance, as an antenna. Simply stated, this argument suggests that the geomagnetic forces between and among the planets indeed play a role in influencing human behavior.

HEMISPHERES OF THE BRAIN

LEFT	RIGHT
Sequential thinking	Intuitive thinking
Language processing	Processing pictures
Words	Face recognition
Logical/technical	Holistic/creative
Parts	Whole

During the DAY When AWAKE	
Sun	
CONSCIOUS	SUBCONSCIOUS
Processing the day's events through words	Emotions and ESP

During the NIGHT when ASLEEP	
	Moon
SUBCONSCIOUS	"CONSCIOUS"
NREM	REM
Thinking type dreams	Bizarre, pictorial dreams

Fig. 12.9. During the day, the left hemisphere predominates because we tend to think in words, and thus it is the more conscious hemisphere. At night, a reversal takes place, and the subconscious right hemisphere comes alive. What was "conscious" becomes "subconscious" and what was "subconscious" becomes "conscious." This is our dreaming mind, what Freud and Jung called the unconscious and what Rampa called the oversoul.

Note the interesting symmetry in this model (see below). The left hemisphere has the Sun at the top, followed by Jupiter and Mars, and, catty-corner, the right hemisphere has the Moon at the bottom, followed by Jupiter and Mars. The bottom quadrant of the left hemisphere is dominated by Mercury, Saturn, and Venus, and the top quadrant of the right hemisphere is dominated by Venus, Saturn, and Mercury. The two hemispheres are identical opposites, except that the Moon replaces the Sun.

ASTROLOGICAL TYPES
Linked to Hemispheres of the Brain

FIRE/Thinking	**AIR/Intuitive**
♌ Leo • ☉ SUN	♎ Libra • ♀ Venus
♐ Sagittarius • ♃ Jupiter	♒ Aquarius • ♄ Saturn
♈ Aries • ♂ Mars	♊ Gemini • ☿ Mercury
EARTH/Sensing	**WATER/Emotional**
♍ Virgo • ☿ Mercury	♏ Scorpio • ♂ Mars
♑ Capricorn • ♄ Saturn	♓ Pisces • ♃ Jupiter
♉ Taurus • ♀ Venus	♋ Cancer • ☽ MOON

Fig. 12.10. The left hemisphere correlates with the thinking and sensing types and their corresponding signs and planets and the right hemisphere correlates with the intuitive and emotional types and their corresponding signs and planets. Note that the key difference is that the left is ruled from the top down (conscious to unconscious) by the Sun and the right is ruled from the bottom up (unconscious to conscious) by the Moon.

The full model suggests that each hemisphere of the brain has a different circadian (day/night cycle) rhythm, the left aligned with the Sun and the right aligned with the Moon. It is well known that the circadian rhythm is monitored by the pineal gland (the crown chakra in kundalini yoga), located in the center of the brain. The pineal gland produces serotonin and melatonin, two neurotransmitters whose molecular structure greatly resembles LSD. As stated earlier, powerful dosages of LSD are measured in micrograms, that is, in the millionth-of-a-gram range. The pineal gland and brain are sensitive to minuscule chemical changes, so much so that the mind is profoundly altered for hours at a time in a cascading effect when LSD is ingested. It is also known that the pineal gland is stimulated by photons, particles of light that enter the brain via the optic nerve, that is, through the eye. In other animals, like deer, changes in the Sun's position trigger the pineal gland to put the female into estrus during mating season.

Continuing with this thought process, just as each hemisphere has a conscious and unconscious side, the "thinking" and "intuitive" types are more conscious and the "sensing" and "emotional" types more subconscious. We therefore can extrapolate and suggest the schemata for a link between the cosmos and hemispheric organization. According to this topology, each sign of the zodiac would correspond to a different hemisphere of the brain. Split into four quadrants, the overall superstructure would be as follows:

LEFT	RIGHT
Conscious	**Subconscious**
THINKING/Conscious	INTUITIVE/Conscious
SENSING/Subconscious	EMOTIONAL/Subconscious

Fig. 12.11. The fourfold division of the two hemispheres of the brain

The left hemisphere is the more "conscious" hemisphere, even for right-brained types, for the simple reason that language is processed in the left hemisphere, and we structure our world, and therefore *think,* in words. Thus, during the day, the left dominates because it is processing the day's events. At night, however, a reversal takes place. What has been "conscious" becomes "unconscious" and what has been "unconscious" becomes "conscious" through dreams. That is why REM is a more right-brained procedure.

The most "conscious" part of the brain is the top quadrant of the left hemisphere. The most intuitive part is the top quadrant of the right hemisphere. Right-brained thinking is intuitive thinking. If a person is one of the fire/thinking signs (Leo, Sagittarius, or Aries), then if a pure type, he or she should be a more left-brained dominant. The most likely channel for psychic ability, according to this model, is the bottom quadrant of the right hemisphere linked to the three water signs (Scorpio, Pisces, and Cancer).

According to this theory, each of the planets is influencing the brain in two ways: 1) by its geomagnetic force, and 2) by its idiosyncratic prismatic reflection of sunlight. I am suggesting, therefore, that for (2) above, reflected sunlight from each of the planets affects the brain in different ways.

It is not my intention to pronounce these astrological speculations as anything other than theories. However, it is worth keeping in mind that

1. Astrology has been studied by almost every major scientist before the modern era (such as Ptolemy, Copernicus, Kepler, Newton, and Galileo).
2. There is empirical evidence supporting aspects of this doctrine (see Gauquelin's studies).
3. Astrology has set up an amazing typology that has influenced many personality theorists.
4. Astrology suggests that humans and the cosmos are intimately linked.

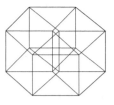

13

ANOTHER LOOK AT
$E = MC^2$

Although Professor Gora passed away over ten years ago, he did have occasion to read through parts of this manuscript. When he read the section on precognition and such synchronicities as the *Titan/Titanic* episode and the unusual number of similarities between the deaths of Lincoln and Kennedy, he said, "It was God playing tricks on us." However, when I showed him an earlier version of this chapter on $E = mc^2$, and my suggestion that if mind is a form of energy, it may dwell in a realm square to the speed of light, he was intrigued, but ultimately saw the idea as "nonsense." "Well, what about this," I said, referring to an article from the *New York Times,* which was about a recalibration of the age of the universe. At the time, which was no doubt in the late 1970s or early 1980s, cosmologists had agreed that the universe was about seven billion years old. This new article stated they now thought it to be about fourteen billion years old. "Where did these extra seven billion years come from?" I insisted. Edwin pushed away the article with a dismissive air. "It's only a factor of two," he said.

Force is nothing other than the principle of change.

G. W. Leibniz

A study of the history of $E = mc^2$ makes it clear that it was derived from the equation for force, which Leibniz conceived in the mid-1600s, approximately 250 years before Einstein: Force = (mass) (velocity)² or $F = mv^2$.

This idea of linking force to change and time came from the ancient philosophers who had lived nearly two thousand years earlier. Leibniz had studied Plato and Aristotle's ideas on form, essence, and existence (e.g., entelechy: the mechanism responsible for life and growth), which, in turn, derived from the Pythagorean view that the study of geometry and music was essential for understanding the construction of the cosmos. Leibniz hypothesized that all that is derives from force. His monad, discussed earlier, is a complex notion. Not only is it a microcosm of the macrocosm; the monad can also be conceived as "a center, which expresses an infinite circumference."[1] Everything is part of the whole.

In Gora's article "Energy, Entropy, and Evolution," he writes, "The word 'energy' had been popular with the ancient Greeks, but more in the sense of what we now call 'action.'" Citing Hermes Trismegistos' *Of Energy and Feeling,* which Gora, no doubt, translated from the original Greek, he wrote, "'Energies, thought in themselves incorporeal, are in bodies, and act through bodies. . . . Things once called into being for some purpose or some cause . . . can never stay inactive of their proper energy. . . . If bodies are on earth, they are subject of dissolution but must serve as places and organs for the energies. The energies, however, are immortal, and the immortal is eternally bodymaking, it is energy.' This sounds like a prophetic vision of Einstein's mass-energy law, the famous $E = mc^2$."[2]

And, in fact, Einstein was not the first to publish his famous equation. Earlier physicists who predated him in this respect include S. Tolver Preston in 1875, Jules-Henri Poincaré in 1903, and Olinto De Pretto and F. Hasenöhrl in 1904.[3] In fact, Einstein's original equation, formulated in 1905, was $L = mv^2$, where L is a form of radiation or energy and v refers to the velocity of light. $E = mc^2$ did not arise until seven years later, when Einstein replaced L with E and v with c.[4] In his book *E = mc²: A Biography of the World's Most Famous Equation,* David Bodanis discusses a question that I have pondered for more than thirty years: Does c^2 have any meaning other than being a mathematical device to help complete the equation?[5] I have come to the conclusion that c^2 indeed does bear some key relationship to the structure of energy and the delineation of a dimension or dimensions that exceed the so-called physical one, which is bound by c, that is, bound by the speed of light.

Let us go back to $F = mv^2$, which was proposed by Leibniz and used by his contemporary and cofounder of calculus, Isaac Newton, in his equation for the total energy of each body, which took into account both kinetic and potential energy.[6] "Why," Bodanis asks, "is the squaring of the velocity of what you measure such an accurate way to describe what happens in nature?"

> One reason is that the very geometry of our world produces square numbers. When you move twice as close toward a reading lamp, the light on the page you're reading doesn't simply get twice as strong. . . . The light's intensity increases four times as much. . . . The interesting thing is that almost *anything* that steadily accumulates will turn out to grow in terms of simple squared numbers. If you accelerate on a road from 20 mph to 80 mph, your speed has gone up by four times. But it won't take merely four times as long to stop if you apply brakes and they lock. Your accumulated energy will have gone up by the square of four, which is sixteen times. That's how much longer your skid will be.[7]

Bodanis goes on to say that "mass is simply the ultimate type of condensed or concentrated energy. Energy is the reverse: it is what billows out as an alternate form of mass under the right circumstances."[8] So in the case of $E = mc^2$, which was simply, albeit brilliantly, an outcropping of Leibniz equation for force, the c^2 component refers to an "enormous conversion factor . . . of how mass gets magnified if it is fully sent across the '=' of the equation."[9]

As I read this, c^2 is thus not just a mathematical prop, but rather has some physical correlate, even though nothing physical can travel faster than c, let alone c^2. The key word here is *travel*. At the turn of the century, Max Planck discovered Planck's constant, h, which describes the size of a quantum of energy, e.g., a light particle (6.626 × 10⁻³⁴ joule/second), which Planck saw as a mathematical constant he needed in order to explain certain calculations concerning electromagnetic effects. Without this tiny additional figure, the calculations would not come out right. Planck saw this measure as a mathematical prop, but, according to Einstein biographer Walter Isaacson, Einstein took the leap and suggested that this number referred to the amount of energy in a wave-packet of light, which a number of years later came to be called the photon.[10] I am suggesting essentially that we do the same thing with c^2.

Consider the problem with photons: Are they waves or are they particles? Bohr and Einstein suggest that they are both. Sometimes they act like a wave and sometimes like a particle. But if indeed they are wave-packets, this would suggest that they cover an area because they are moving laterally as well as unidirectionally. So photons proceed forward at the speed of light, but each packet could also be seen as moving at right angles to itself as a contained wave, c in one direction and c at right angles to that direction as this packet continues on its way. That's $c \times c = c^2$, the area contained in the wave packet. This can be seen as a field equivalent to c^2. [11]

And if the photon is three dimensional, energy would be moving at the speed of light in three dimensions, on a graph, the x, y, and z axes, and thus c^3, at least within the confined space of the photon. Look, for instance, at the relationship of the diameter of a circle as compared to its circumference. It is the fraction 22/7 or π (pi). But if you divide 7 into 22, you get a number with no end, pi = 3.142159265 . . . Does this not indeed, at least in a symbolic way, pair well with

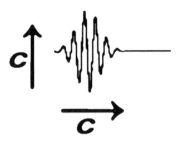

Fig. 13.1. c² wave-packet

the dual properties of the photon? Where 22/7 is an exact amount (quantum), the decimal goes on forever (wave). At the same time, this "wavicle" is also spinning, perhaps like the electron at 1.37 times the speed of light. Just like any gyroscope, such a high torque will produce a precession or transmission of energy at right angles to the direction of spin. Just as Ouspensky and other mathematicians have shown and we have discussed, such a process suggests the introduction of a new dimension at right angles to the existing one.

Where Leibniz calculated the amount of energy added to a system in motion, that is, acceleration, Einstein's equation takes into account the energy already existing in the body before it moves. This can be called, or is linked to, its inertial mass, the mass already present, which includes not only its atomic structure and weight but also the complex movement of the Earth, solar system, and galaxy, all of which are part of its inherent inertial mass. For example, the Earth is imbued not only with its own inertial mass spinning at a rate of 1,000 mph, but also with the inertial mass of its movement around the Sun, 60,000 mph, and with the solar system's movement around the Milky Way, at 489,000 mph, and the Milky Way itself is hurtling through space at an astonishing 1,300,000 mph (which is nearly 2 percent of the speed of light)! Clearly, this is an enormous amount of imbued energy.

But mass, even though it is equivalent to energy, is not the same thing as energy. "Specifically, matter has energy and mass, whereas fields only have energy." Where matter concentrates a good deal of energy, fields concentrate a much smaller level. Looking at it this way, the mind can be seen as a "field" that is generated by the matter of the brain.

The equation $E = mc^2$ says outright that E is operating in some way in a realm square to the speed of light. It is not "traveling" faster than the speed of light. It is operating/existing in a realm square to the speed of light. Since mind is some form of neuroelectric or field energy, it follows that mind may also operate in a dimension square to the speed of light. Another way to look at this is from simply a geometric point of view. Even if we restrict "mind" to be a complex electrical field occurring inside the brain, it operates in a space that comprises photons stretching out in at least three dimensions: in the height, width, and depth of the brain, that is, the size of the brain. Thus, if mind is a form of energy, then $E \approx \text{Mind} = mc^2$, with c^2 or higher powers thereof correlating to some type of photonic energy in the mind/brain field extending into at least these three dimensions. Perhaps this could be expressed in a way analogous to Avagadro's number, which refers to "the amount of protons in a gram of pure protons,"[12] but in terms of electromagnetic energy, as, for example, the amount of energy equivalent to mental energy as found in an iPod.

Fig. 13.2. Albert Einstein

According to Duncan Copp, in his article "Einstein's Dream," "[S]pecial relativity binds together not three but *four* dimensions, three for space (forward-backward, left-right, up-down) and the fourth being time."[13] He further notes that "whichever of these dimensions we are moving through, the sum total of the velocities of all of them must *always* be equal to the speed of light."[14] Since space together with time must equal the speed of light, Cobb points out that if a body travels at a speed close to *c,* then the time component is reduced to nil. Since photons themselves travel at the speed of light, time does not exist for them. Photons are "as young as they were on the day they were created during the big bang billions of years ago. Time waits for no man, unless you can travel very close to the speed of light."[15]

But I contend that our minds are already traveling at or beyond the speed of light. Support for this hypothesis can be seen in the fact that the brain itself, as a set of organized organic molecules, works on an electrical alternating current. It is well known that different brain-wave AC frequencies, measured by the EEG in cycles per second (cps), are associated with different states of consciousness. The following chart catalogs each brain wave state we go through when falling asleep.

BRAIN WAVES: WAKING, SLEEPING, AND DREAMING

State	EEG	Attributes
β BETA	14–60+ cps	Waking
α ALPHA	7–14 cps	Relaxed
ø THETA	3–7 cps	Very drowsy
Δ DELTA	1–3 cps	Out cold, asleep
β REM	14–60+ cps	Vivid dreams, beta state
α ALPHA	7–14 cps	NREM
ø THETA	3–7 cps	NREM
Δ DELTA	1–3 cps	NREM
β REM	14–60+ cps	Vivid dreams, beta state

Because brain waves are electrically based, they are already dwelling in a realm isosynchronous to the speed of light. Further, different mental states are thus directly related to corresponding specific AC frequencies. Note also that the REM state occurs when the individual is awake and also when he is asleep, when beta is associated with rapid eye movements, REM. In a previous chapter, I have hypothesized that this difference is most likely associated with the different functions of each hemisphere: beta when conscious would be controlled by the left hemisphere and during REM, when dreaming, by the right hemisphere.

According to Tom Bearden, there are two types of photons: scalar photons, which are related to mental activity, and longitudinal photons, which are more associated with physical things. Thus, "when a given electrodynamic mind and its dynamics are coupled to a living body, there is a great coherent 'pairing' of the scalar photons of the mind with the longitudinal photons of the 3-d space physical body (brain and nervous system). But to have that coupling at all, there must exist jillions of 'voltage spikes' that represent the mind to body coupling." This, according to Bearden, is achieved via the "jillions" of dendrite connections in the brain where occurs this interface of mental and physical electrodynamic realms.[16]

Fig. 13.3. *Calvin travels faster than the speed of light. (From* The Authoritative Calvin and Hobbes, *by Bill Watterson.)*

MOMENTUM

Einstein used the notation c^4 in his original equation, which is commonly reduced to $E = mc^2$. His full equation before it was simplified is as follows: $E = m^2c^4 + p^2c^2$, where p^2c^2 refers to momentum, or mass in motion. One way or another, Einstein had considered, at least in a mathematical sense, higher-order representations of the speed of light.

According to Bertrand Russell, "[M]ass is only a form of energy, and there is no reason why matter should not be dissolved into other forms of energy. It is energy, not [the] matter that is fundamental in physics."[17] Energy is more akin to the idea of "field" than just "energy."[18]

Informational components and negentropic or teleological aspects permeate all matter. There is order to the structure of the universe. If mind, in an evolutionary sense, at the high end of this process, exists in a three-dimensional oscillating electronic brain where $E = mc^2$ is associated with our mental field, then mind, as a form of energy, may travel "faster" than the speed of light, that is, in a c^2 realm. Mind, if simply conceived as an electronic process, may never operate as *slow* as the speed of light. The world of thought operates at a higher rate of vibration than does matter. That is why the mental realm is not bound to the laws ascribed to the physical world. Mind exists in a dimension beyond the physical dimension, which is limited by light speed. For the experiential mind, time really does not exist. The electronic dimension in which it lives should be thought of in more spatial terms involving field properties and dynamic action, rather than in linear, physical terms. Mind is more a spatial concept, or, if you like, a hyperspatial concept.

Mind sees the whole and thus the simultaneity of all of existence, and it can project itself in imagination backward and forward into time. As a holographic monad, or microcosm of the macrocosm, it has "a center, which expresses an infinite circumference" (Leibniz). It thus transcends, at least conceptually, this thing called time.

BEYOND LIGHT SPEED

I finally ascertained with a reasonable degree of certitude, and to my amazement, that the sun was at a constant positive potential of about 216,000,000,000 volts. Thus, the secret of the cosmic rays was revealed. Owing to its immense charge, the sun imparts to minute positively electrified particles prodigious velocities, which are governed only by the ratio between the quantity of free electricity carried by the particles and their mass, some attaining a speed exceeding fifty times that of light.

NIKOLA TESLA

Consider, for a moment, the Andromeda galaxy. We can look in a telescope and see our closest galactic neighbor. It is roughly the same size and shape as the Milky Way, approximately 100,000 light-years long and 30,000 light-years across at its center bulge. Clearly the galaxy is one cosmic entity. As a whole, it spins, and all of its parts/stars spin with it. All are connected to one another via some inertial/scalar or gravitational force. When seen as a gestalt, that is, as a single complex

Fig. 13.4. Spiral galaxy

entity, it is simply absurd to view it in terms of the speed of light. In fact, from our vantage point, we can see the whole thing at once, the entire 100,000-light-year span that reaches us as a single spinning unit. Andromeda is about 2.3 million light-years away, which means we see it as it was 2.3 million years ago.

One theoretician who stands out as looking at the structure of the universe in a way different from conventional physicists is Nobel Prize–winner Hannes Alfven, who thinks that great magnetic and plasma fields are just as important as gravity as prime movers and shakers in the creation of galaxies, solar systems, and planets. Alfven states, "Galaxies are formed by giant vortex filaments carrying electric currents towards the galactic center."[19]

Clearly, such huge rotating electromagnetic plasma fields are held together by a collective force that easily subsumes the speed of light. They extend out, in a sense, instantaneously, much like an octopus with tendrils, thousands, even tens of thousands of light-years long, giant spiraling force fields that shape and energize these massive spinning star systems. Alfven suggests that his plasma model is better suited for explaining the shapes and locations of galaxies and general texture of the universe. Since this idea may supersede the big bang theory, Alfven is more a proponent of the steady state theory with a universe that has no beginning and no end.

Frankly, I see merit in both views, because on the one hand, considering the immense size, complexity, and age of the universe, it seems presumptuous to think that we really know that the universe began with a big bang. Therefore, the idea that the universe always *was* has appeal. On the other hand, the mystical idea that the universe began when the One reflected upon him/herself to split up into the multitudes is very attractive. Also, there is Hubble's finding that the universe is expanding, which lends itself better to a big bang theory rather than one of steady state.

Looking at it on a more local level, our Sun is ninety-three million miles away. Dividing 186,000 miles/second into that figure, we can calculate that it takes photons between eight and nine minutes to reach the Earth. But is there any doubt that our solar system is one unit? Aren't we/the Earth connected to the Sun in an intimate way that has nothing to do with the speed of light? In an analogous sense, how long would it take an ant to crawl from a person's toe to his head? From the ant's point of view, it might take nine seconds. But from the person's point of view, the toe and head are simultaneously connected. The Andromeda galaxy is one entity. There must be a number of mechanisms that connect one end of this galaxy to its other end simultaneously. Humans measure the distance in light-years, the time it takes light traveling at 186,000 miles/second to go in one year (roughly 6 trillion miles, or 6×10^{12}), but from the point of view of the galaxy, this is barely a nanosecond in relationship to the nonlocal timescale that the galactic entity is operating on. If looked at from the speed-of-light point of view (6 trillion miles/year $\times$ 100,000 years = 6×10^{17} miles), information would have to travel about c^3 to traverse the galaxy in an instant. Clearly, there are physical mechanisms in the universe that easily transcend the speed of light. The angular momentum and/or plasma field that shapes and springs galaxies is at least one. As we have seen, the etheric sea that it floats in is probably another.

Nonlocal, tachyonic, or hyperspatial realms and angular momentum are more than a few ways to portray a higher-order spacetime scale that can allow one side of a galaxy to be immediately in touch with its other side. Higher powers of c may be another way to explain simultaneous interconnections within solar systems, galaxies, local galaxy clusters, and the universe.

If mind is energy and $E = mc^2$, then the theory of relativity would place the mind out of any time dimension because time ceases to have meaning as it approaches the speed of light, let alone that of light squared. For Einstein, time slows down and then ceases to exist when the speed of light is reached. What he didn't realize, however, when he came across an instance of infinity by using the Lorenz transformation (see page 297), as Ouspensky points out, it was really a discovery of a different dimension, one that exists in the c to c^2 realm. Yet paradoxically, consciousness can exist only in "the Now." We can perceive only in the present. If the mind was bound by the physical properties of matter, perception or self-perception could not take place because the mind must be able to move in a way that is unaffected by the physical inertia of the system. The ability to *observe,* by its nature, must be in a dimension unaffected by the onslaught of time. It must be able to cogitate and also access memories. From the neurophysiological perspective, this involves information retrieval stored in the atomic structure of the neurons of the brain, specifically in the protein chains at the ends of the dendrites.

It is interesting to consider Lisa Randall's hypothesis that if gravity is ubiquitous, it will "fall off exponentially" as it enters or interfaces with a higher dimension.[20] The implication here is be that as one changes from the physical to the mental, the function of gravity recedes "exponentially," and if so, this would correlate nicely to the idea of matter being in one dimension and spirit being in another.

The galaxy, solar system, and orthorotating Earth have a momentum that is essentially unalterable, predetermined, locked in by the physical properties of this complex spinning system. We hurl along prescribed pathways. Time is linked to the precise, ongoing, ever-changing *present* moment of this dynamic activity. The Earth, Sun, and galaxy have no real freedom, no say in their destiny. They go where they are forced to go. In order for change to take place, there must be some mechanism that transcends this onward march of time. That something, at least for us, is our imagination, our mind, which can only operate in the ongoing Now, yet can continually transcend this Now by its very nature.

Nuclear explosions not withstanding, physicists have pretty much ignored the c^2 aspect of Einstein's equation (let alone his use of c^4). In fact, a nuclear explosion is not explained as involving something that is traveling square to the speed of light. For physicists, c^2 is simply a mathematical aspect of the equation with no material counterpart. It is seen as a conversion factor used to change mass to energy.[21] However, c^2 may indeed have some corresponding hyperphysical, tangible counterpart in the same way that v^2 in the equation $F = mv^2$ refers to the physical act of acceleration. If v^2 has a physical counterpart, c^2 must as well. Einstein's equation suggests there is an equivalence among mass, energy, and light; to that I am adding thought. It is the powers of the velocity of light, c, c^2, c^4, that delineate the interpenetrating dimensions of the universe. We know from the Lorentz transformation that no physical object can reach the speed of light:

$$m_1 = \frac{m_0}{\sqrt{1-v^2/c^2}}$$

where m_0 = old mass
$\quad m_1$ = new mass
$\quad v$ = velocity
$\quad c$ = speed of light

Looking at the denominator, as v (velocity) approaches c (speed of light), v^2/c^2 approaches the integer 1. Subtracting this figure from 1 and taking its square root, we can see that the denominator begins to approach 0 (zero). Thus, the numerator, m_0, begins to approach infinity (because it is being divided by a smaller and smaller amount). This standard equation explains why nothing can travel as fast as the speed of light: If it did, it would obtain infinite mass. This, of course, is impossible. This same equation holds true if we replace m with t. Time, t_1, would have no meaning if light speed were achieved, as it would obtain an infinite quantity. That was Einstein's finding.

But suppose the mind never operates as *slowly* as the speed of light because it exists in a dimension square to the speed of light. Physicists have a hard time explaining how a person could speculate about the future because the future does not dwell in physical time, but it lives easily in the world of thought or realm of imagination. We have already shown that the elementary particles that make up the brain, including the atoms at the ends of the dendrites that house our memories, are already spinning in excess of the speed of light (i.e., $1.37 \times c$). Our mind is not bound by the speed of light because it transcends time all of the time.

Thought is energy. But the act of thinking is also the creation (redistribution or transcription) of energy, and therefore may be considered a fundamental force of the universe. Thought, which we know is associated with electronic states already operating at light speed, is a force transcribed into action by the brain. A man designs a house and builds it. A mental event becomes physical. Energy in the c^2 or higher realm steps down into a physical world bound by c, the speed of light, via the processes of intention and willpower, through the mechanism of the brain.

Consider the following redistribution of Einstein's equation:

$$E/m = c^2$$

Energy divided by mass equals the speed of light squared. This chapter is suggesting that this equation has a physical or hyperphysical counterpart and demarcates a different dimension, one that operates in some way outside the realm of time.

We know from major theories of the unconscious that certain aspects of the unconscious do not deal with time. We have all experienced the feeling that time does not exist. This perception is valid. It is most vivid in dreams, especially when we return to a dream world that we may have been in previously, whether it was six months or fifteen years ago. In a key aspect of human perception, time has no meaning.

Taking a little liberty with Einstein's equation and extrapolating, we come up with the following:

Thought ≈ Electronic Energy ≈ Mind
Mass ≈ Matter

Energy/Mass ≈ Mind/Matter ≈ c²
Mind Over Matter ≈ the Speed of Light Squared

The mind, as well as other forms of "pure" energy, exists in a "beyond time" dimension simply because time has no meaning for this realm. We can watch our bodies grow old and decay over time, all the while knowing that our "me-ness" is ageless, and somehow transcends the aging process of the body. It is our physical body that exists in the physical spacetime continuum, whereas our mind performs in a dimension at least square to the physical dimension. Said in another way, thought, which may be seen as a special form of organized light, exists in a different dimension from the one we call the physical.

If there was a whole range of consciousness above matter, what was this consciousness like? Since it was conscious, it clearly would have to be aware of itself. Would it then be spirit? That seemed likely. Would this spirit have a tangible form? That seemed likely too. After all, form was only an energy field, and consciousness was certainly energy. Also, a higher consciousness seemed likely to have a greater energy field, just as lighter hydrogen released more energy than heavier uranium. Then this thought hit me, I saw man's dilemma. Only two ways to go—descent into matter (gravity) or ascent into spirit (consciousness)—for matter was in a process of devolving—compacting, growing heavier—subject to the forces of Time; and spirit was in a process of evolving— expanding, growing lighter, subject to the forces of Space.

U. S. ANDERSON

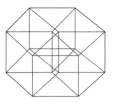

14
FINAL THOUGHTS

While working on the final draft of this book, I came upon notes in preparation for the work dating all the way back to 1972, some thirty-six years ago. They were written by a twenty-five-year-old taken from some unknown book on physics. One line at the top of one of the pages reads as follows:

> What is the basic framework of the universe that allows us to determine the direction of the gyroscope and say it does not change? I know this was something that has fascinated me since I was in the ninth grade. How is it that a gyroscope can set up its own coordinates in space irrespective of the movement of the Earth? Clearly, for this gyroscope, there is a frame of reference that supersedes the movement of the Earth, and possibly the solar system and maybe even the galaxy. There is something fundamental to orthorotation with respect to the fabric of space-time.

In 1972/73, while working on my master's thesis at the University of Chicago, I came upon serious experimental studies conducted by J. B. Rhine and many others on man's higher abilities including telepathy, out-of-body experiences, psychokinesis, and even communicating

300

with the dead. These were all topics that, before then, I had no particular interest in. My research beyond the worlds of Freud and Jung led me, ultimately, to F. W. H. Myers's concept of the Universal Mind: we are all interconnected, including the souls of those of the living with those who have passed away. Who among us has not had a dream of a departed relative or friend that does not at all seem like a wish fulfillment or fantasy, but feels instead like a real and true communication? I remember distinctly sitting in my dorm room shaken because what I had thought was reality was no longer so.

Once I had entered this world, I had a number of paranormal experiences, some of which have been mentioned throughout the text. In 1978, I was at a New Age conference where one of the speakers was Matthew Manning. Author of *The Link,* Manning claimed to be a clairvoyant who was in touch with various well-known dead artists who could do their drawings through him. Essentially, they would take over his hand. His book is peppered with many of these fine drawings. During one of his demonstrations—I think he was drawing a Picasso— I saw a distinct ring of light above his head, a halo. It was not the only time I would see a halo.

Several years later, I was at a conference in Providence with a Tibetan Buddhist monk. Dressed in a suit and tie, the monk was unable to speak English. He would speak a sentence and it would get translated, and so on. After his talk, he sat down and began to chant. As unbelievable as this may sound, the speaker transformed before my eyes into a monk dressed in orange robes. I tapped my friend on the shoulder to ask if he saw the same thing, and he thought I was teasing him. The speaker stopped his chant and he appeared again to me in Western attire. The monk chanted again and I saw it again! This experience of transfiguration was just about the strangest thing that has ever happened to me.[1]

About this same time I was able to interview Andrija Puharich, the psychic researcher and medical doctor who had contributed so much to this field. He was my father's age, probably about sixty. I met him at his house in Ossining, New York, with my partner, Howard Smukler, editor of the *Journal of Occult Studies* and *ESP Magazine*. I had met him before. He was just a normal fellow, in casual attire, with his hair a little bit of a mess, certainly not dressed to impress two young men who came to interview him for an article. It was at this time that Puharich told us that everything in the room we were sitting in had levitated except the piano, and that he had a watch whose dials moved when he received messages from extraterrestrials. Here was a medical doctor with more than fifty patents, many concerned with perfecting the hearing aid. At the same time, he was a successful author and scientist who studied such topics as psychic surgery, paranormal metal bending with Uri Geller, and telepathy with the well-known psychic and founder of the Parapsychology Foundation, Eileen Garrett. These last studies had been conducted in the early 1950s with John Hayes Hammond Jr., inventor of radio guidance systems at Hammond's compound in Gloucester, Massachusetts. Hammond and Puharich had placed Eileen Garrett in a Faraday cage, which screens out electromagnetic waves, and yet she still achieved significant results in telepathy experiments, thus suggesting that ESP is a force or means of information exchange that lies outside the normal electromagnetic range. This was Puharich, looking for scientific answers to a complex and far-out field, and he was also working with the U.S. military.

Because of what had happened to me, having had my consensus reality shattered, I asked Puharich how he dealt with this strange world he had entered. He told me that there was no need to set some hard-and-fast rules as to what was and what was not true. "Reality is a learning curve," Puharich said. And that idea has stuck with me. I didn't need to decide for sure if he had a screw loose. I allowed myself to simply be open to the possibility that the world is a lot more complicated than we think. Was Puharich in touch with extraterrestrials? I don't know, but what I do know is that he did credible studies in the field of telepathy, and that he didn't just say that Uri Geller had psychic ability, but rather set up Geller to do experiments at seventeen laboratories around the world including such places as Stanford Research Institute and Lawrence Livermore Laboratories. Concerning extraterrestrials, I know, or at least believe, that humans cannot be the smartest entities in the cosmos. There has to be an intelligence hierarchy of some type, and we may be near the top but we certainly cannot be the top. But, then, intelligence is a difficult thing to define. If a leaf can convert sunlight, dirt, and water into food, is it smarter than a human? If a lizard can regenerate a tail, is that not also a very advanced form of intelligence? Look what our immune system does every day. And what about DNA? Gurdjieff says that one of the functions of humans is to help the cosmos evolve. That is certainly an interesting idea compatible with and predating the physicist's notion of the Anthropic Principle, which holds that the universe was set up in a way that allowed humans, as observers of the universe, *to* evolve.

I tried in this book to rattle a few cages. Certainly I have great respect for Einstein and his achievements. However, what is clear is that Einstein did not operate isolated in a box. Rather, he was heavily influenced by predecessors, particularly the abstract mathematician Hermann Minkowski, who introduced to Einstein the concept of the *imaginary* number i (the square root of negative one) to better explain his model for a *physical* universe. What I came to realize, particularly after studying Ouspensky and his idea to change the Einsteinian four-dimensional spacetime continuum into one of six dimensions, was that Einstein's model was incomplete. His four-dimensional model could not explain the simple idea of thinking about, let alone planning for, the future.

It also always seemed self-evident that the generally accepted notion that space is empty was untenable—an ether of sorts must exist. There is no emptiness in space. Place a Hubble telescope anywhere. Every point in space has the intersecting light from every available star, and thus each point codes for every other point. Although mainstream physicists have dismissed the idea of an ether (or renamed the nineteenth-century concept so that it is almost unrecognizable, e.g., the

Higgs boson), this was not the case with Einstein, the theoretician they kept misinterpreting. Einstein *knew* there was an ether and, as quoted earlier, agreed with Lorenz on this point, memorialized in a letter to Lorenz, and Einstein also lectured on the topic at Leiden. But ultimately, he essentially ignored ether theory, he ignored consciousness as a force in and of itself, and he rejected Weyl's suggestion that at least one more dimension was necessary better to explain how to tie in gravity with the other three forces, a goal Einstein never achieved. Further, he ignored aspects of his theory that were inherently contradictory. Nothing could travel as fast as the speed of light, yet matter itself was held together by photons, which did indeed operate in a realm that was at least as fast as the speed of light.

Having taught courses in the field of consciousness studies for over thirty years, I have been privy to information that has been so casually dismissed by mainstream scientists. Forget such far-out things as psychokinesis, UFOs, precognition, and psychic metal bending; mainstream scientists will not accept even the simple notion that humans have telepathic abilities. This puts me in an odd position of knowing something that was, and still is, clearly being rejected by the people whom we rely on for understanding and modeling the universe. Yet at the same time, these individuals, mostly theoretical physicists, will speak of wormholes, black holes, the big bang, superstrings, and parallel universes as if these ideas are actual things rather than simple theories, and as if these speculations are somehow more acceptable than the idea that one person could dream about what another person was doing or thinking.

It really is time for university neuroscientists and quantum physicists to use the MRI, for example, and begin to study the phenomenon of telepathy. It tends to occur between twins and between people with similar interests. It also tends to occur quite often through dreams, that is, during the REM cycle. If we can understand the mechanism of telepathy, it may very well help unlock more information toward the goal of understanding the nature of our consciousness, the fundamental forces of the universe, how they interact and their potentiality.

I have done my best to place the ideas herein within a historical framework. Consciousness as a force, in and of itself, can no longer be ignored. If the work shakes the cobwebs and causes a few bright lights to revisit the nature of this post-Einsteinian world, then I think *Transcending the Speed of Light* has succeeded. I'd like to end the with a quote from one of the key inspirations of the book.

> *The goal of science should be to understand the basic principles of how our presently perceived cosmos came into being, how it is being maintained, how it may be transformed, and how . . . out of it . . . we may transform ourselves.*[2]
>
> CHARLES MUSÈS FROM
> *CONSCIOUSNESS AND REALITY*

NOTES

PREFACE

1. Louise B. Young, *The Unfinished Universe* (New York: Touchstone Books, 1987), 204.

CHAPTER 1. CONSCIOUSNESS AND THE ANTHROPIC PRINCIPLE

1. Paul Tobin, *Is the Universe "Fine-Tuned" for Life?* www.geocities.com, 2000.
2. Cited in Richard Morris, *The Edges of Science* (New York: Simon & Schuster, 1990), 212.
3. T. H. Leahey, *A History of Psychology* (Englewood Cliffs, N.J.: Prentice Hall, 2000), 144–45.
4. Amit Goswami, *The Self-Aware Universe* (New York: Tarcher/Penguin, 1993), 190.
5. J. G. Bennett, *Gurdjieff: Making a New World* (Santa Fe, N. Mex.: Bennett Books, 1993), 146–47.
6. Edwin Gora, "Pythagorean Trends in Modern Physics," part 2 (Kingston, R.I.: *MetaScience Quarterly,* 1985), unpublished.
7. A physicist of the old school, Gora knew both Arnold Sommerfeld and Werner Heisenberg, who kept the half-Polish, half-German young Edwin Gora from the Gestapo during World War II; they continued their friendship after the war.

8. Werner Heisenberg, cited in Gora, "Pythagorean Trends in Modern Physics," 4.

9. J. Wheeler, cited in the foreword in John Barrow and Frank Tippler, *The Anthropic Cosmological Principle* (Oxford, England: Clarendon Press, 1986), vii.

10. Henry Stapp, *Mind, Matter and Quantum Mechanics* (New York: Springer, 1993), 234.

11. Dreaming as a separate state of consciousness involves its own attributes, such as primary-process language, separate and distinct from the waking state of consciousness. Dreams often transcribe emotions into pictorial symbols.

12. Herbert Read, *Origins of Form in Art* (New York: Horizon Press, 1965), 156.

13. Gary Lachman, *A Secret History of Consciousness* (Great Barrington, Mass.: Lindisfarne Books, 2003), 22–26; Henri Bergson, *Mind-Energy* (London: The Macmillan Company, 1920).

14. Marc Seifer, *Wizard: The Life and Times of Nikola Tesla* (New York: Citadel Press, 1997), 202–203; Nikola Tesla, "The Problem of Increasing Human Energy," *The Century Magazine* (June 1901): 177–211.

15. R. S. Jones, *Physics as Metaphor* (New York: New American Library, 1982).

16. Ed Fredkin, cited in an interview with Robert Wright, "Did the Universe Just Happen?" *Atlantic Monthly* (April 1988): 29–42.

17. Freeman Dyson, *Disturbing the Universe* (New York: Harper & Row, 1981), 250.

18. David Chalmers, "The Puzzle of Conscious Experience," *Scientific American* (December 1995): 80–86.

19. Alfred Taylor, "Meaning and Matter," in *Consciousness and Reality,* ed. Arthur Young and Charles Musès (New York: Outerbridge/Dutton, 1972), 172.

20. It is this author's belief that this transference occurs in mammals during the REM stage of sleep. The animal must be split off from conscious activity, that is, be at rest, so the information accumulated during the day can be transferred to the memory banks. According to this theory, this occurs during the REM state. It takes a tremendous amount of energy to be awake and conscious. After this information is absorbed during the day, it must be processed and cataloged and transferred from some short-term holding tank associated with mRNA to long-term banks, which we know are found in new formations on the dendrites.

21. Taylor, "Meaning and Matter," 174–75.

22. A word about intelligent design: Before Darwin, the prevailing theory of evolution came from Lamarck: use and disuse. A giraffe got a long neck because it used it; the human appendix was disappearing because it wasn't being used. Darwin writes in *On the Origin of Species* that he *does* indeed accept Lamarck's theory in a limited way. For instance, he mentions that domestic ducks have weaker bones than their wilder counterparts, because the wild ducks fly south for the winter. Darwin suggests that the more dominant mechanism stems from the ability/necessity for an animal to adapt to the environment. Those that adapt, survive, those that don't, die out. To account for incremental changes, Darwin adopts the idea of chance mutation and survival of the fittest.

What Darwin is really doing is portraying evolution from the perspective of regarding the environment as the key defining factor. In an urban setting, for example, squirrels are better suited to survive than are bobcats, bison, and porcupines. Darwin, however, is missing the negentropic (goal-directed) aspect, and the fact that squirrels haven't changed for tens of thousands of years. What is it within the animal, within all of life, that propels one species or another to make a go of it? Henri Bergson called this the élan vital, or zest for life.

The intelligent design argument starts with a false premise. Obviously, animals evolve. What fun would it be if God had the whole thing planned out? As my sister-in-law, puts it, "He waters the garden and watches it grow." Gurdjieff calls this involution, or "help from above." Ultimately, life could be seen as an inner calling to absorb sunlight in all its derivative forms, to preserve the self, and to propagate. Any full theory of evolution must take into account *first cause,* the inner motivation of an organism to survive, as well as the outer constraints imposed by the environment.

23. P. D. Ouspensky, *In Search of the Miraculous* (New York: Harcourt & Brace, 1960), 14.

24. Chalmers, "The Puzzle of Conscious Experience," 80–86.

25. Cited in Eric Kandel, *In Search of Memory: The Emergence of the New Science of Mind* (New York: W. W. Norton, 2006), 382.

26. Liotti, et al., "Brain Responses Associated with Consciousness and Air Hunger," *PNAS* (February 13, 2001), 2035–2040.

27. Rudolf Steiner, *An Outline of Occult Science* (New York: Anthroposophic Press, 1972), 34.

CHAPTER 2.
PARAPSYCHOLOGY AND ESOTERIC THOUGHT

1. James Coates, *Photographing the Invisible* (London: L. N. Fowler & Co., 1911), iv.

2. Eli Beers, *Mind as the Cause and Cure of Disease* (Chicago: Self-published, 1914), 3.

3. Ibid., 83.

4. Holger Hyden, "RNA: A Functional Characteristic of the Neuron and Its Glia," *Brain Functions,* ed. Mary Brazier (Berkeley: University of California Press, 1964), 29–60.

5. Beers, *Mind as the Cause and Cure of Disease,* 84.

6. Carl Jung, "The Psychological Foundation of Belief in Spirits," *Psychology and ESP,* ed. Robert Van Over (New York: Mentor Books, 1972), 92–108.

7. Ibid.

8. Beers, *Mind as the Cause and Cure of Disease,* 84.

9. Kate Boheme, *Realization Made Easy* (Holyoke, Mass: Elizabeth Town Co., 1917), 37.

10. Ibid., 38.

11. Beers, *Mind as the Cause and Cure of Disease,* 181.

12. Carl Jung, *Memories, Dreams, Reflections* (New York: Vintage Press, 1963), 150.

13. Richard Ingalese, *History and Power of Mind* (New York: Occult Book Concern, 1901).

14. Ibid., 71.

15. Ibid., 73.

16. Ouspensky, *In Search of the Miraculous.*

17. E. H. Anderson, *Psychical Developments* (Toledo, Ohio: n.p., 1901).

18. Ibid.

19. Ibid.

20. Ibid.

21. Ingalese, *Mind as the Cause and Cure of Disease,* 47.

22. Ibid.

23. Frederich R. Marvin, M.D., *The Philosophy of Spiritualism and The Pathology and Treatment of Mediomania* (New York: Asa Butts & Co., 1874), 16–17.

24. Leonard Landis, M.D., *Psychoanalysis and Beyond* (New York: American Association of Independent Physicians, 1924), 133–34.

25. Ibid., 172.

26. Ibid., 187.

27. Ibid., 200.

28. Camille Flammarion, *Stories of Infinity, Lumen—History of a Comet in Infinity* (Boston: Roberts Brothers, 1873), 7–8.

CHAPTER 3.
TOWARD A PHYSICS OF CONSCIOUSNESS

1. John White, ed., *Frontiers of Consciousness* (New York: Avon, 1974), 313.

2. Ken Wilber, *Spectrum of Consciousness* (Wheaton, Ill.: Quest, 1977), 35.

3. See Uri Geller's description of his own teleportation in *My Story* (New York: Warner Books, 1976), 264.

4. E. Harris Walker, *Psychic Explorations,* eds. Edgar Mitchell and John White (New York: G. P. Putnam & Sons, 1974), 544–68.

5. Helen Graham, "Color Therapy Then & Now," www.Innerself.com, 2005.

6. Michael Nyberg, "Killing Mosquitoes with Sound," www.ctsciencefair.org/news, 2001.

7. Barry Lynes, *The Cancer Conspiracy: Betrayal, Collusion, and the Suppression of Alternative Cancer Treatments* (Delmar, N.Y.: Elsmere Press, 2000), 69.

8. Ibid., 246.

CHAPTER 4. THE BIRTH OF THE NEW PHYSICS

1. Newton, cited in Arthur Koestler, *The Sleepwalkers* (New York: Random House, 1959), 503.

2. Maxwell, cited in L. Williams, *The Origins of Field Theory* (New York: Random House, 1966), 130–131.

3. Ibid., 131.

4. Ibid., 135.

5. Ouspensky, *In Search of the Miraculous,* 356.

6. Herbert Dingle, foreword to *Duration and Simultaneity* (New York: Bobbs-Merrill, 1965), xx.

7. L. Williams, *The Origins of Field Theory,* 61.

8. Ibid., 62.

9. Ibid., 116.

10. Lew Price and B. Herbert Gibson, "Is There a Dynamic Ether? A New Reality for 21st-Century Physics," www.promedia.net/users/greenbo, 2007.

11. G. Lombardi, "The Michaelson-Morley Experiment," www.drphysics.com, 1997.

12. Alpheus Smith and J. Cooper, *Elements of Physics* (New York: McGraw Hill, 1975), 98.

13. Ibid.

14. Edwin Gora, "Pythagorean Trends in Modern Physics," part 1 (Kingston, R.I.: *MetaScience Quarterly*, 1979, 1983), unpublished.

15. Nikola Tesla, *The Inventions, Researches and Writings of Nikola Tesla* (Belgrade: Tesla Museum, 1956), L149.

16. Einstein, cited in Ronald Clark, *Einstein: The Life and Times* (New York: World Publishing Co., 1971), 78.

17. Ibid., 33.

18. Gora, personal discussions, 1990.

19. Clark, *Einstein: The Life and Times,* 122.

20. Charles Musès and Arthur Young, *Consciousness and Reality* (New York: Avon Books, 1972), 111.

21. Ibid., 121.

22. By 1920, Einstein had become so famous for his theory of relativity that pressure was put on the Nobel Prize committee to honor him. However, according to Einstein biographer Isaacson, because the theory was controversial, the committee gave Einstein the prize for another finding, his discovery of the photoelectric effect.

23. Einstein, cited in Clark, *Einstein: The Life and Times,* 122.

24. Minkowski, cited in Clark, *Einstein: The Life and Times,* 123.

25. Clark, *Einstein: The Life and Times,* 124.

26. Musès and Young, *Consciousness and Reality,* 452–53.

27. Ibid., 453.

28. Ibid., 111.

29. Ibid., 130.

30. George Gamow, *Thirty Years That Shook Physics* (New York: Anchor Press, 1966), 46.

31. Ibid., 69.

32. Ibid., 119–20.

33. Ibid., 125–26.

34. Ibid., 129–30.

35. Gora, "Pythagorean Trends in Modern Physics," 208.

36. Sharla Stewart, "How to Catch a Higgs," *University of Chicago Magazine* (2001): 21.

37. Adapted from R. Feynman, *Five Easy Pieces* (New York: Audiotape lectures, 1961).

38. Musès and Young, *Consciousness and Reality,* 448.

39. Ibid., 449.

40. Ibid., 140.

41. Brant Finstad, "Superstrings: The New Quantum Particle," 2003.

42. Einstein, "Ideas and Problems of the Theory of Relativity," Nobel Prize lecture, July 11, 1923.

43. Michael Atiyah, "Hermann Weyl, Biographical Memories," www.stills.nap.edu, 2003.

44. Ibid.

45. Einstein to Kaluza, April 21, 1919, in Walter Isaacson, *Einstein: His Life and Universe* (New York: Simon & Schuster, 2007), 338.

46. Ibid.

47. Andrew Revkin, "Lisa Randall, Physicist" (*Rolling Stone,* November 15, 2007), 158.

48. Richard Morris, 222.

49. Mayeul Arminjon, "Lorentz-Poincaré Relativity and a Scalar Theory of Gravitation," http://geo.hmg.inpg.fr, 2007.

50. Musès and Young, *Consciousness and Reality,* 122.

51. Isaacson, 312–13.

52. P. D. Ouspensky, *New Model of the Universe* (New York: Random House, 1971), 343.

53. Ibid., 346.

54. Ibid., 348.

55. Ibid., 352.

56. Ibid., 353.

57. Ibid., 360.

58. Ibid., 227.

59. Ian McCausland, "Anomalies in the History of Relativity," *Journal of Scientific Exploration* 13, no. 2 (1999): 271–90.

60. Nikola Tesla, cited in *Tesla Said,* edited by John Ratzlaff (Milbrae, Calif.: Tesla Book Company, 1984), 240–42.

61. Ibid., 361.

62. Oupensky, *New Model of the Universe,* 358. The idea that "light has weight" was established by Professor Lebedeff of Moscow: "[L]ight when falling on bodies produces a mechanical pressure on them. . . . Lebedeff's discovery was very important for astronomy; for instance, it explained certain phenomena, which had been observed at the passing of the tail of a comet near the sun. . . . [and] is supplied a further confirmation of the unity of the structure of radiant energy."

63. Tesla, cited in *Tesla Said,* 362.

64. Ibid.

65. Ibid., 364.

66. Ibid., 364–69.

67. Henri Bergson, *Duration and Simultaneity* (New York: Bobbs-Merrill, 1965), vi.

68. Ibid., vi–vii.

69. Ibid.

70. Ibid., viii.

71. Paul Anderson, "Four Talks on Tim" (regarding Gurdjieff's teachings), (Gurdjieff Electronic Publishing, www.gurdjieff.org, 2003).

72. Bergson, *Duration and Simultaneity,* viii.

73. Ibid., 163.

74. Ibid., 64.

75. Ibid., xxxv–xxxvi.

76. Dingle, foreword to *Duration and Simultaneity,* xxxii.

77. Isaacson, *Einstein: His Life and Universe,* 107.

78. Ibid.

79. Ibid., 108.

80. James Coleman, *Relativity for the Layman* (New York: Penguin, 1958), 42.

81. Ibid., 43.

82. Ibid., 45–46.

83. Ibid., 51–54.

84. Isaacson, *Einstein: His Life and Universe,* 107.

85. Ibid., 62.

86. Ibid., 78.

87. Ibid., 6, 125.

88. Robert S. Fritzius, "Abbreviated Biographical Sketch of Walter Ritz," www.datasync.com, 2006.

89. Bjorn Overbye, "Warped Minds, Bent Truths," *Nexus* 14, nos. 5–6 (2007).

90. McCausland, "Anomalies in the History of Relativity," 271.

91. Clark, *Einstein: The Life and Times,* 328–29.

92. Isaacson, *Einstein: His Life and Universe,* 125.

93. Ibid., 318.

94. Ibid.

95. Ibid.

96. Ibid.

97. Ibid., 319.

98. Ibid.

99. Ibid.

100. Ibid., 316–17.

101. Clark, *Einstein: The Life and Times.*

102. Ibid., 639–41.

103. Isaacson, *Einstein: His Life and Universe,* 327.

104. Orrin Dunlop, *Radio's 100 Men of Science* (New York: Harper & Row, 1944), 156–58.

105. T. C. Martin, *The Inventions, Researches, and Writings of Nikola Tesla* (New York: Electrical Experimenter Publisher, 1893), 149.

106. Seifer, *Wizard: The Life and Times of Nikola Tesla,* 103, 464, 499.

107. Tesla, Leiden lecture, 1920.

CHAPTER 5. ETHER THEORY REVISITED

1. Overbye, "Warped Minds, Bent Truths," 1.

2. David Wilcox, *Convergence VIII: Scientific Proof of the Nature of a Multi-Dimensional Harmonic Universe,* www.ascension200.com, 2000.

3. Mikhail Shapkin, "Unknown Manuscript of Nikola Tesla," farshores.org/wmtesla.htm.

4. Wilcox, *Convergence VIII.*

5. Nikola Tesla, "Tesla on the Peary North Pole Expedition," *New York Sun,* 1905, cited in *Tesla Said,* edited by John Ratzlaff, 90–91.

6. Nikola Tesla, "World System of Wireless Transmission of Energy," *Telegraph & Telephone Age,* October 16, 1927.

7. Nikola Tesla, "Pioneer Radio Engineer Gives View on Power," *New York Herald Tribune,* September 11, 1932, in *Tesla Said,* ed. Ratzlaff, 240–42.

8. Nikola Tesla, "Nikola Tesla Tells of New Radio Theories," *New York Herald Tribune,* September 22, 1929, 1, 29, in *Solutions to Tesla's Secrets,* edited by John Ratzlaff (Milbrae, Calif.: Tesla Book Company, 1981), 225–26.

9. Nikola Tesla, "Dr. Tesla Writes of Various Theories of His Discovery," *New York Times,* February 6, 1932, 16, cited in ibidem, 237–38.

10. Nikola Tesla, "Tesla Sees Evidence Radio and Light Are Sound," *New York Times,* April 8, 1934, 9, C. 1, cited in ibidem, 258–60.

11. Joseph Alsop Jr., "Beam to Kill Army at 200 Miles, Tesla Claims," *New York Herald Tribune,* July 11, 1934, 1, 15, cited in ibidem, 110–14.

12. Tesla, "Nikola Tesla Tells of New Radio Theories," ibidem, 225–26.

13. Tesla, cited in *Tesla Said.*

14. Ibid.

15. Fritjof Capra, *The Tao of Physics* (Berkeley, Calif.: Shambhala Press, 1975), 210.

16. Edwin Gora, "Letter to the Editor," *Journal of Occult Studies* 2 (1978): 207–8.

17. Ibid.

18. Gora, private correspondence, 1978.

19. Marc Seifer and H. Smukler, "The Puharich Interview," *Gnostica* (September 1978), 46.

20. Capra, *The Tao of Physics,* 64.

21. Ibid., 208–9.

22. Isaacson, *Einstein: His Life and Universe,* 218.

23. Gora, "Pythagorean Trends in Modern Physics," part 2 (unpublished).

24. The ether is "omnidirectional" and "the very basis of space-time." Warren York, *Scalar Technology,* www.teslatech.info, 2007.

25. Ibid.

26. Wright, *Did the Universe Just Happen?,* 38.

27. Albert Einstein, quoted in a letter to Hermann Weyl, May 26, 1923, regarding his search for a unified field theory, in Isaacson, 351–52.

28. Albert Einstein, "On The Method of Theoretical Physics," the Herbert Spencer lecture, Oxford, June 10, 1933.

29. Nikola Tesla, quoted in *My Inventions*, edited by Ben Johnston (Williston, Vt.: Hart Brothers, 1981), 61. Cited in M. Seifer, *Wizard*, 22.

30. Marc Seifer, *Wizard: The Life and Times of Nikola Tesla*, 121.

31. V. Bujic, *Magnetic Vortex, Hyper-Ionization Device*, www. linux-host.org/energy/magvid.htm, 2006.

32. Tesla, cited in *Tesla Said*, 22–23.

33. Ron Hatch, http://egtphysics.net, 2003.

34. Wilcox, *Convergence VIII*, 2.

35. Gamow, *Thirty Years That Shook Physics*, 2–3.

36. Nikola Tesla, "The Eternal Source of Energy of the Universe: Origin and Intensity of Cosmic Rays," *Tesliana* (Belgrade: Tesla Museum, 1932), 56–59.

37. S. Uchii, "Mach's Principle," Kyoto University, www.bun.Kyoto-u.ac.jp, 2001.

38. Bujic, *Magnetic Vortex, Hyper-Ionization Device*, 1–2.

39. York, *Scalar Technology*.

40. Isaacson, *Einstein: His Life and Universe*, 147.

41. Ibid.

42. Albert Einstein, "Ether and the Theory of Relativity," University of Leiden lecture, May 5, 1920.

43. Peter Weiss, "Jiggling the Cosmic Ooze," *Science News* 159 (March 10, 2001): 153.

44. Meg Urry, "The Secrets of Dark Energy," *Parade* (May 27, 2007): 4–5.

45. Marcia Bartusiak, *Through a Universe Darkly* (New York: Avon, 1993), 216.

CHAPTER 6. ETHER THEORY: THE MENTAL ASPECT

1. Kepler noticed that the time it takes a planet to circle the Sun squared divided by the distance from the planet to the Sun cubed was the same for all planets. This is harmonic law $P^2/D^3 = K$, where P = the period of revolution around the Sun, D = distance to the Sun, and K = a constant. He also noticed that a planet will sweep out equal areas between itself and the Sun in equal times even though it moves at varying speeds and distances (depending upon where it is in its orbit). P^2/D^3 led to Newton's law of gravity: $m_1 m_2/D^2 = K$, where m_1 = mass of one body and m_2 the mass of another, D = the distance between them, K = a constant.

It should also be noted that Kepler's discovery of equal areas being swept out under a curve led Newton to his formulation of calculus. Einstein comes in because he noticed that P^2/D^3 did not work for the planet Mercury. Einstein realized that Mercury's speed would be a factor. Before his interest, the anomaly was basically ignored. This helped him prove his theory of relativity.

2. Roemer had noticed that eclipses of Jupiter's moons differed depending on where the Earth was in its orbit around the Sun. When it was farther from the Sun, the eclipses were found to occur at a later time. This could only be attributed to the possibility that light had a finite speed and took longer to reach the Earth at that time. The differences in the distance compared to the different times of the eclipses enabled him to calculate light speed.

3. Arthur Koestler, *The Sleepwalkers,* 503, quoting Newton: "That gravity should be innate . . . and essential to matter, so that one body may act upon another, at a distance through a vacuum, without the mediation of anything else . . . is to me so great an absurdity . . . but whether this agent be material or immaterial, I have left to the consideration of my readers."

4. Mach's principle that the influence of all the stars is causally related to the stability of matter in a local system derives directly from Newton's own words: "If a vessel hung by a long cord is so often turned about that the cord is strongly twisted, then filled with water and held at rest together with the water thereupon by the sudden action of another force it is whirled about the contrary way . . . the surface of the water will at first be plain, as before the bucket began to move . . . [eventually, however] the swifter the motion becomes, the higher will the water rise. . . . This ascent of the water shows its endeavor to recede from the axis of motion. See *Ernst Mach, His Work, Life and Influence,* by John T. Blackmore (Berkeley: University of California Press, 1972), 101–2, 248.

5. Edwin Gora states that interconnectedness does not necessarily imply information exchange, although I argue that it does.

6. Brendon O'Regan quoting David Bohm in *Psychic Explorations*, edited by John White and Edgar Mitchell (New York: G. P. Putnam & Sons, 1974), 46.

7. Kathy Wallard, "Why Our Planet and Galaxy Take Us for a Spin," *Newsday* (March 9, 2004).

8. Dane Rudhyar, *The Sun Is Also a Star* (New York: Dutton, 1975), 24.

9. Sarfatti, *Space-Time and Beyond*, 281.

10. Whitrow, *The Nature of Time* (New York: Penguin, 1980), 359.

11. Adapted from G. J. Whitrow, 353; Jack Sarfatti, 281.

12. Sarfatti, *Space–Time and Beyond*, 234.

13. Ibid., 289.

14. Whitrow, *The Nature of Time*, 360.

15. Ibid., 361.

16. Ouspensky, *New Model of the Universe*, 354.

17. Ibid., 354–56.

18. Ibid., 396.

19. William Lyne, *Occult Ether Physics* (Lamy, N. Mex.: Creatopia Productions, 1997), 37.

20. John Hasted, David Bohm, Edward Bastin, and Brendon O'Regan, "Experiments in Psychokinetic Phenomena" in *The Geller Papers*, edited by Charles Panati (Boston: Houghton Mifflin, 1976), 183–96.

21. Scott Reyburn, "Uri Buys Spoons in $3.8 Million Savoy Sale at Boham's," www.bloomberg.com, 2007.

22. Marc Seifer, *Speculations on the Nature of the Mind* (Kingston, R.I.: unpublished, 1976).

23. Ibid.

24. Jonathan Margolis, *Uri Geller: Magician or Mystic?* (London: Orion Publishing Group, 1999), 265.

25. Clark, *Einstein: The Life and Times*, 122–25.

26. Robert Wright, *Did the Universe Just Happen?* 38.

27. Musès and Young, *Consciousness and Reality*, 454, 462–63.

28. Gamow, *Thirty Years That Shook Physics*, 123–24.

29. Musès and Young, *Consciousness and Reality*, 130.

30. Ibid., 455.

31. Coleman, *Relativity for the Layman*, 76.

32. Bergson, *Duration and Simultaneity*, 49.

33. Roger Penrose, *Shadows of the Mind: A Search for the Missing Science of Consciousness* (London: Oxford University Press, 1994), 384.

34. Bergson, *Duration and Simultaneity*, 49.

35. Fred Alan Wolf, *Star Wave: Mind, Consciousness and Quantum Physics* (New York: Macmillan, 1984), 24–25.

36. Ibid.

37. Goswami, *The Self-Aware Universe,* 190.

38. Charles Musès, interviewed by Jeffrey Mishlove (from *Roots of Consciousness* online, circa 1990), www.WilliamJames.com. Updated at www.mishlove.com.

39. Rodney Collins, *Theory of Celestial Influence* (London: Vincent Stuart, 1970), 23.

40. Madame Blavatsky, *The Secret Doctrine* (Wheaton, Ill.: Quest Books, 1973), opening paragraphs.

41. Ibid.

42. It is interesting to note that the molecular structure of the neurotransmitters (e.g., adrenaline, serotonin, acetylcholine, dopamine) contains a carbon ring. Is it possible that this elegant geometric *shape* has a vibratory rate conducive for carrying and transferring cerebral information?

43. Louis Acker, "Mind: A Holographic Computer," unpublished, 1978.

44. Ken Wilber, *Spectrum of Consciousness* (Wheaton, Ill.: Quest Books, 1977), 200.

45. Whitrow, *The Nature of Time,* 364–65.

46. *Bulletin of Atomic Scientists,* vol. XXXI, 9 (November, 1975).

47. Arthur Koestler, *Roots of Coincidence* (New York: Vintage Press, 1972), 63.

48. Lyne, *Occult Ether Physics,* 11–12.

49. Ibid., 13.

50. Rudhyar, *The Sun Is Also a Star,* 24.

51. Ibid.

CHAPTER 7. COEVOLUTION OF SCIENCE AND SPIRIT

1. Werner Heisenberg, *Physics and Beyond* (New York: Harper & Row, 1971), 227.

2. In his 1980 book, *Wholeness and the Implicate Order* (London: Routledge & Kegan Paul), David Bohm wrote, "There is the germ of a new notion of order here. This order is not to be understood solely in terms of a regular arrangement of objects (e.g., in rows) or as a regular arrangement of events (e.g., in a series). Rather, a total order is contained, in some implicit sense, in each region of space and time. Now, the word 'implicit' is based on the verb 'to implicate.' This means 'to fold inward' . . . so we may be led to

explore the notion that in some sense each region contains a total structure 'enfolded' within it" (p. 149).

3. Arthur Koestler, cited in a speech by Pir Vilayat Khan.

CHAPTER 8. MONAD OF MIND

1. Alexander Luria, *The Working Brain: An Introduction to Neuropsychology*
(New York: Basic Books, 1973), 31.

2. Holger Hyden, cited in Mary Brazier, ed., *Brain Functions* (Berkeley:
University of California Press, 1964).

3. Kandel, *In Search of Memory*, 383.

CHAPTER 9. SYNCHRONICITY

1. Arthur Koestler, *Case of the Midwife Toad* (New York: Random House,
1971), 138.

2. Ibid.

3. Ibid., 139.

4. Ibid., 140.

5. Carl Jung, *Portable Jung*, edited by J. Campbell (New York: Viking Press,
1971), 506.

6. Ibid., 517–18.

7. Dennis A. Williams, "The Energy Tangle," *Newsweek* (April 16, 1979):
72.

8. Ira Progoff, *Jung, Synchronicity and Human Destiny* (New York: Julian
Press, 1973), 145.

9. Jung, *Portable Jung*, 518.

10. D. Marks and R. Kammann, *The Psychology of the Psychic* (Buffalo, N.Y.:
Prometheus Books, 1980), 170.

11. I met the Amazing Randi one evening at the Larchwood Inn, in Wakefield,
Rhode Island, in 1981. A complete skeptic, he had absolutely no belief
in synchronicity. Unbeknownst to either of us, we were both flying out
of town two days later to two different destinations. As a total surprise,
for the first leg of the trip, we both had to take the same plane and were
assigned seats next to each other. During the trip Randi continued to
argue that there was nothing to synchronicity!

12. Alan Vaughan, "Synchronicity," *Psychic Magazine* (August 1975): 17.

13. B. Walsh, M. Seifer, and H. Smukler, "Synchronicity and Seriality as a Partial Explanation of the Large Number of Oil Spills in New England," *Journal of Occult Studies* 1 (1977): 76–84.

14. Within a month of the Lennon meeting, and to my utter surprise, I met Uri Geller for the first time. It was in his apartment as I was writing a lead story on him for *ESP Magazine*. Naturally, I recounted the entire Lennon tale. Later, Uri told me a very odd story about Lennon. They had become good friends and had exchanged gifts. Uri had bent a spoon for John, and John had given Uri a shiny round object the size of a golf ball he said came from an extraterrestrial that had entered the bedroom in his apartment at the Dakota, near where I had met him, in New York City.

    Sure, this story is impossible to believe, but it underscores one of the points of this book, namely that we humans know very little. We have blinded ourselves to the wider reality. Why is it that we can so easily accept extraterrestrials in motion pictures like *The Day the Earth Stood Still, Close Encounters, E.T.,* and *Galaxy Quest* but vehemently reject the possibility that these fictional excursions may actually be based on truth? One day in 1978, I asked Andrija Puharich how he was able to deal with the far-out world he had entered. He had supposedly seen levitation, miraculous healing, contacted ETs via a special watch, and witnessed so-called Geller children, who like Uri, could bend metal psychically. Puharich said that life was a learning curve and that he didn't need to decide black or white what reality is.

15. Coincidentally, my father was born in Toronto, where my partner, Tim Eaton, who now lives in California, grew up. As it turns out, eighty years ago, my grandfather Isadore Seifer worked for the Eaton Clothing Store, which is a very successful enterprise in Toronto. When I asked Tim about this store, he informed me that the owners are cousins of his.

16. C. Holden, "Twins Reunited," *Science* 80 (November 1980): 55–59.

17. Ibid., 57.

18. Ibid., 55–59.

19. Edward Bruce Taub-Bynum, "Psi, Dreams and the Family Unconscious," *MetaScience Quarterly* 4 (1987): 29–45.

20. S. Dennis, Bioelectromagnetics Society, San Antonio meeting, *Planetary Association for Clean Energy Newsletter* (1981): 2, 10; M. A. Persinger,

"ELF Meditation in Spontaneous Psi Events," *Psychoenergetic Systems* 3 (1979): 155–69.

21. Persinger, "ELF Mediation in Spontaneous Psi Events," 156.

22. Ibid.

23. Mishlove, *Roots of Consciousness.*

24. It is well known that telepathic phenomena transcend distance, can pierce Faraday cages, which screen out electromagnetic frequencies, and so on. Thus, on one level, these kinds of synchronistic phenomena may be "beyond space and time." Nevertheless, one way or another, at some point these occurrences must step themselves down to electromagnetic frequencies that are compatible with normal cognitive and brain-wave functions.

CHAPTER 10. PRECOGNITION AND THE STRUCTURE OF TIME

1. Quoted in H. J. Forman, *The Story of Prophecy* (Toronto: Farrar & Rhinehart, 1936).

2. Ibid.

3. C. Panati, *Supersenses* (New York: Quadrangle Books, 1974), 216–17.

4. S. Edmunson, *Miracles of the Mind* (Chicago: Charles Thomas Publishers, 1965); I. Stevenson, "Precognition of Disasters," *Journal of the American Society for Psychical Research* 2 (1970); Douglas Dean, "Precognition and Retrocognition," *Psychic Explorations,* edited by Mitchell and White (New York: G. P. Putnam & Sons, 1974), 164.

5. Dean, "Precognition and Retrocognition," 164.

6. Vaughan, "Synchronicity," 56–61.

7. Albert Einstein, *Essays on Science* (New York: Philosophical Library, 1934), 49.

8. J. B. Rhine, *New World of the Mind* (New York: William Sloan Associates, 1953), 27.

9. Ibid., 67.

10. Ibid., 27.

11. J. W. Dunne, *An Experiment in Time* (London: Faber & Faber, 1929), 50.

12. Ibid., 55–56.

13. Ibid., 2.

14. Whitrow, *The Nature of Time,* 364.

15. Ibid., 365.

16. H. F. Saltmarsh, *Foreknowledge* (London: G. Bell & Sons, 1938), 87–88.

17. Ibid., 1–2.

18. Ibid., 5.

19. Ibid., 26.

20. Ibid., 80.

21. Ibid., 87.

22. Ibid., 100.

23. Ibid., 98.

24. Ibid., 100.

25. Lobsang Rampa's "hall of future probabilities" in *Feeding the Flame* (London: Corgi Books, 1971), 94–96; Rampa, *Wisdom of the Ancients* (London: Corgi Books, 1974), 9.

26. Whitrow, *The Nature of Time,* 368.

27. Ouspensky, *New Model of the Universe,* 71.

28. C. A. Bragdon, *A Primer of Higher Space* (Tucson: Omen Press, 1972); R. Rucker, *Geometry, Relativity and the Fourth Dimension* (New York: Dover Publications, 1977), 2.

29. Ouspensky, *New Model of the Universe,* 72.

30. Ibid., 81.

31. Ibid., 82.

32. Ibid., 83.

33. Ibid., 86.

34. O. Whicher, *Projective Geometry* (London: Steiner Press, 1971), 241–51.

35. Dunne, *An Experiment in Time;* Saltmarsh, *Foreknowledge;* Ouspensky, *New Model of the Universe.*

CHAPTER 11. SIX-DIMENSIONAL UNIVERSE

1. Ouspensky, *New Model of the Universe,* 374.

2. Ibid., 375.

3. Ibid., 397.

4. Ibid.

5. Ibid., 397–98.

6. Ibid., 400.

7. K. I. T. Richardson, *The Gyroscope Applied* (New York: Philosophical Library, 1954).

8. Musès and Young, *Consciousness and Reality*, 453–54.

9. Roger Penrose, "On the Origin of Twister Theory," *Gravitation and Geometry*, www.users.ox.ac.uk, 2007, 40–49.

10. Ibid.

11. Ibid., 46.

12. Ibid.

13. Ibid., 47.

14. Ibid., 48.

15. Wright, *Did the Universe Just Happen?*, 34.

CHAPTER 12. PATTERNS OF PROPHECY

1. Lobsang Rampa, *Wisdom of the Ancients*.

2. Progoff, *Jung, Synchronicity and Human Destiny*.

3. Vaughan, *Patterns of Prophecy*, 219.

4. Krippner in Vaughan, *Patterns of Prophecy*.

5. Jung, cited in Progoff, *Jung, Synchronicity and Human Destiny*.

6. Progoff, *Jung, Synchronicity and Human Destiny*, 80–82.

7. Forman, *The Story of Phrophecy*, 3.

8. Dane Rudhyar, *The Astrology of Personality* (Garden City, N.Y.: Doubleday, 1970).

9. A word about Pluto: It's round, it circles the sun, it has at least one moon—it's a planet. In fact, because Chiron, Pluto's moon, is so large, this system could almost be called a binary planet. The attempt by early-twenty-first-century astronomers to strip Pluto of its rightful status is an amazing and rather sad example of groupthink.

10. Michel Gauquelin, *Cosmic Clocks* (New York: Regnery Co., 1967).

11. www.astro.com/mtp/mtp64e.htm.

CHAPTER 13. ANOTHER LOOK AT $E = MC^2$

1. Velimir Abramovic, "Inadequacy and Inconsistency as Imperfection in the Differential Calculus," www.Scienceoftime.org, 2005.

2. Edwin Gora, "Energy, Entropy, and Evolution" (September 25, 1978), an unpublished class handout.

3. Overbye, "Warped Minds, Bent Truths."

4. David Bodanis, *E = MC²: A Biography of the World's Most Famous Equation* (New York: Walker Publishers, 2000).

5. Ibid.

6. Gora, "Energy, Entrophy, and Evolution." Where "kinetic energy . . . depends on the location of all the surrounding bodies. $E = E_{kin} + E_{pot} = mv^2/2 + V(x,y,z)$. The kinetic energy $[E_{kin}]$ is one half of the product of the mass times velocity squared while the potential energy $[E_{pot}]$ depends on the location of all the surrounding bodies—it depends upon the location of a body in space $(x,y,z$—Cartesian coordinates in a given reference system)." Cleary, this idea of linking the energy of a body to all the surrounding bodies is the basis of Mach's principle.

7. Bodanis, $E = MC^2$, 67–68.

8. Ibid., 69.

9. Ibid.

10. Francisco Flores, "The Equivalence of Mass and Energy," www.calpoly.edu, 2004.

11. Isaacson, *Einstein: His Life and Universe.*

12. www.chem.uwimona.edu.jm:1104/courses/pH/avono.html.

13. Duncan Copp, in his article "Einstein's Dream," Channel 4 Science Microsite, 2003.

14. Ibid.

15. Ibid.

16. Tom Bearden, www.cheniere.org.

17. Russell, *Human Knowledge,* 291.

18. Flores, "The Equivalence of Mass and Energy."

19. Eric Lerner, "The Big Bang Never Happened," *Discover* (June 1988): 72–78.

20. Lisa Randall, *Warped Passages* (New York: HarperCollins, 2006). Taken from her own reading of her book on CSPAN.

21. Richard Rhodes, *The Making of the Atomic Bomb* (New York: Simon & Schuster, 1986), 260.

CHAPTER 14. FINAL THOUGHTS

1. Hypnosis seems to me to be the most logical explanation of this experience. Through his chant, the monk caused a change in my state of consciousness. However, if hypnosis is the cause of the hallucination—I was in a normal state of consciousness during this talk—then it would also seem that telepathy might play a role, for if the monk wanted to give some type of command, he was doing it in Tibetan, a language I certainly do not understand. So my guess is that the monk's chant created a telepathic link triggering an idiosyncratic eidetic image in my mind. Concerning the halo, auras and corresponding energy manifestations are fairly easy to see, once one begins to look for them. Aside from the aura, I have seen small bursts of light around the head of some students who have stood up in front of a screen during aura experiments; a double aura, the second one known as the etheric, around the head and body of Moshe Dayan; a yellow aura above the shoulders of healer Olga Warrell; and a virtual helmet of light surrounding the head of clairvoyant Elwood Babbitt.

2. Musès and Young, *Consciousness and Reality,* 122.

BIBLIOGRAPHY

Abramovic, Velimir. "Inadequacy and Inconsistency as Imperfection in the Differential Calculus." www.Scienceoftime.org, 2005.

Acker, Louis. "Mind: A Holographic Computer." Unpublished, 1978.

Atiyah, Michael. "Hermann Weyl, Biographical Memories." www.stills.nap.edu, 2003.

Alsop, Joseph, Jr. "Beam to Kill Army at 200 Miles, Tesla Claims." *Solutions to Tesla's Secrets*. Edited by John Ratzlaff. Milbrae, Calif.: Tesla Book Co., 1981.

Anderson, E. H. *Psychical Developments*. Toledo, Ohio: n.p., 1901.

Anderson, Paul. "Four Talks on Tim." Gurdjieff Electronic Publishing, www.gurdjieff.org, 2003.

Anderson, U. S. *Greatest Power in the Universe*. Los Angeles: Atlantic Press, 1972.

Arminjon, Mayeul. "Lorentz-Poincaré Relativity and a Scalar Theory of Gravitation." http://geo.hmg.inpg.fr, 2007.

Barnett, Lincoln. *The Universe and Dr. Einstein*. New York: Time Inc., 1948.

Barrow, John, and Frank Tipler. *The Anthropic Cosmological Principle*. Oxford, England: Clarendon Press, 1986.

Bartusiak, Marcia. *Through a Universe Darkly*. New York: Avon, 1993.

Bearden, Thomas. *Reflections of a 6th Stage Eye: The One Human Problem*. Unpublished manuscript circa 1979, Harvard lecture 1977, and personal communications.

Becker, R. O. "An Application of Direct Current Neural Systems to Psychic Phenomena." *Psychoenergetic Systems* (March–April 1978): 189–97.

Beers, Eli. *Mind as a Cause and Cure of Disease.* Chicago: Published by author, 1914.

Bennett, J. G. *Gurdjieff: Making a New World.* Santa Fe, N. Mex.: Bennett Books, 1993.

Bentov, Itzhak. *Stalking the Wild Pendulum.* New York: Dutton, 1977.

Bergson, Henri. *Duration and Simultaneity.* New York: Bobbs-Merrill, 1965.

————. *Mind-Energy.* London: The Macmillan Company, 1920.

Bethell, Tom. "Rethinking Relativity." www.gravitywarpdrive.com, 2007.

Bird, Christopher. "Interview of Arthur Young." *Psychic* (June 1976).

Blavatsky, Madame. *The Secret Doctrine.* Wheaton, Ill.: Quest Books, 1973.

Bodanis, David. *E = MC²: Biography of the World's Most Famous Equation.* New York: Walker Publishers, 2000.

Boheme, Kate. *Realization Made Easy.* Holyoke, Mass.: Elizabeth Town Co., 1917.

Bohm, David. *Wholeness and the Implicate Order.* London: Routledge & Kegan Paul, 1980.

Bragdon, C. A. *A Primer of Higher Space.* Tucson: Omen Press, 1972.

Bujic, V. *Magnetic Vortex, Hyper-Ionization Device.* www.linux-host.org/energy/magvid.htm, 2006.

Burr, H. S. *The Fields of Life.* New York: World Publishing Co., 1971.

Capra, Fritjof. *The Tao of Physics.* Berkeley, Calif.: Shambhala Press, 1975.

Cassidy, D. B. "What Is Antimatter?" www.cmr.wsu.edu, 2005.

Chalmers, David. "The Puzzle of Conscious Experience." *Scientific American* (December 1995): 80–86.

Cheiro. *Language of the Hand.* New York: TransAtlantic Publishers, 1898.

Clark, Ronald. *Einstein: The Life and Times.* New York: World Publishing Co., 1971.

Coates, James. *Photographing the Invisible.* London: L. N. Fowler & Co., 1911.

Coleman, James. *Relativity for the Layman.* New York: Penguin, 1958.

Collins, Rodney. *Theory of Celestial Influence.* London: Vincent Stuart, 1970.

Copp, Duncan. "Einstein's Dream." www.channel4.com, 2003.

Correa, Paula, and Alexandra Correa. "Consequences of the Null Result of the Michelson-Morley Experiment." *Infinite Energy* 7:38 (2001): 47–64.

Darwin, Charles. *On the Origin of Species.* New York: Gramercy, 1895.

Dean, Douglas. "Precognition and Retrocognition." In *Psychic Explorations.* Edited by E. Mitchell and J. White. New York: G. P. Putnam & Sons, 1974.

Dennis, S. Bioelectromagnetics Society. San Antonio meeting. *Planetary Association for Clean Energy Newsletter* (1981): 2, 10.

Dingle, Herbert. Foreword to *Duration and Simultaneity*. New York: Bobbs-Merrill, 1965.

Dunlop, Orrin. *Radio's 100 Men of Science*. New York: Harper & Row, 1944.

Dunne, J. W. *An Experiment in Time*. London: Faber & Faber, 1929.

———. *Serial Universe*. London: Faber & Faber, 1934.

Dyson, Freeman. *Disturbing the Universe*. New York: Harper & Row, 1981.

Edmunson, S. *Miracles of the Mind*. Chicago: Charles Thomas Publishers, 1965.

Einstein, A. "Ether and the Theory of Relativity." Leiden University lecture, May 5, 1920. www-groups.dcs.st-and.ac.uk.

———. *Essays on Science*. New York: Philosophical Library, 1934.

———. "Ideas and Problems of the Theory of Relativity." Nobel Prize lecture, July 11, 1923.

———. "On the Method of Theoretical Physics." The Herbert Spencer lecture, Oxford, England, June 10, 1933.

Einstein, A., and L. Infield. *The Evolution of Physics*. New York: Simon & Schuster, 1938.

Ellenberger, Henry. *Discovery of the Unconscious*. New York: Basic Books, 1970.

Emery, G. "Sure It's Seasonal Work, But . . ." *Providence Journal, Sunday Magazine* (January 6, 1981): 22–23 (feature article on M. Seifer).

Everett, H. *The Many-Worlds Interpretation of Quantum Mechanics*. Princeton, N.J.: Princeton University Press, 1973.

Faraday, Ann. *Dream Power*. Berkeley, Calif.: Berkeley Publishing Group, 1972.

Feeley, T. "Holography: State of the Art." *MetaScience Quarterly* 1 (1979): 69–77.

Feynman, R. *Five Easy Pieces*. New York: Audiotape lectures, 1961.

Finstad, Brant. "Superstrings: The New Quantum Particle." 2003.

Flammarion, Camille. *Stories of Infinity, Lumen—History of a Comet in Infinity*. Boston: Roberts Brothers, 1873.

Flores, Francisco. "The Equivalence of Mass and Energy." www.calpoly.edu, 2004.

Forman, H. J. *The Story of Prophecy*. Toronto: Farrar & Rhinehart, 1936.

Freud, Sigmund. *Collected Works*. New York: Modern Library, Random House, 1938.

Fritzius, Robert S. "Abbreviated Biographical Sketch of Walter Ritz." www .datasync.com, 2006.

Gamow, George. *Thirty Years That Shook Physics*. New York: Anchor Press, 1966.

Gauquelin, Michel. *The Cosmic Clocks*. New York: Henry Regnery Co., 1967.

Gazzaniga, Michael. *The Bisected Brain*. New York: Meredith Corp., 1970.

Geller, Uri. *My Story*. New York: Warner Books, 1976.

Gora, Edwin. "Energy, Entropy, and Evolution." September 25, 1978, unpublished.

———. "Letter to the Editor, re: Dirac." *Journal of Occult Studies* 2 (1978): 207–208.

———. "Pythagorean Trends in Modern Physics." Part 1. Kingston, R.I.: *MetaScience Quarterly* 2 (1979): 53–72.

———. "Pythagorean Trends in Modern Physics." Part 2. Kingston, R.I.: MetaScience Publications, 1985, unpublished.

Goswami, Amit. *The Self-Aware Universe*. New York: Tarcher/Penguin, 1993.

Graham, Helen. "Color Therapy Then & Now." www.Innerself.com, 2005.

Gregory, A. "Psychological Aspects of Paranormal Phenomena." *Psychoenergetic Systems* 3 (1979): 195–227.

Grossman, M. *Textbook of Physiological Psychology*. New York: John Wiley & Sons, 1968.

Hall, Calvin, and Gardner Lindzey. *Theories of Personality*. New York: John Wiley, 1970.

Hasted, John, David Bohm, Edward Bastin, and Brendon O'Regan. "Experiments in Psychokinetic Phenomena." *The Geller Papers*. Edited by Charles Panati. Boston: Houghton Mifflin, 1976.

Heisenberg, Werner. "Uber quantentheoretische Umdeutung kinematischer und mechanischer Beziehungen." *Z. Physik* 33 (1925): 879.

———. *Physics and Beyond*. New York: Harper & Row, 1971.

Holden, C. "Twins Reunited." *Science* 80 (November 1980): 55–59.

Hudson, Thomson Jay. *The Law of Psychic Phenomena*. Chicago: A. C. McClurg & Co., 1895.

Hyden, Holger. "RNA: A Functional Characteristic of the Neuron and Its Glia." In *Brain Functions*. Edited by Mary Brazier. Berkeley: University of California Press, 1964.

Ingalese, Richard. *History and Power of Mind*. New York: Occult Book Concern, 1901.

Isaacson, Walter. *Einstein: His Life and Universe*. New York: Simon & Schuster, 2007.

Jacobson, Jerry. *The Secret of Life: Perspective in Science*. New York: Philosophical Library, 1985.

Jones, Marc Edmund. *How to Learn Astrology*. Garden City, N.Y.: Doubleday, 1970.

Jones, R. S. *Physics as Metaphor*. New York: New American Library, 1982.

Jouvet, M. "Biogenic Amines and States of Sleep." *Science* 1 (1969): 62–64.

Jung, Carl. *Memories, Dreams, Reflections*. New York: Vintage Press, 1963.

———. *Portable Jung*. Edited by J. Campbell. New York: Viking Press, 1971.

———. "On Synchronicity." In *Portable Jung*. Edited by J. Campbell. New York: Viking Press, 1971.

———. "The Psychological Foundation of Belief in Spirits." In *Psychology and ESP*. Edited by Robert Van Over. New York: Mentor Books, 1972.

Kammerer, Paul. *Das Gets der Serie*. Stuttgart: Deutsche Verlags-Anstahlt, 1919.

Kandel, Eric. *In Search of Memory: The Emergence of the New Science of Mind*. New York: W. W. Norton, 2006.

Koestler, Arthur. *Case of the Midwife Toad*. New York: Random House, 1971.

———. *Roots of Coincidence*. New York: Vintage Press, 1972.

———. *The Sleepwalkers*. New York: Random House, 1959.

Krippner, Stanley. *Song of the Siren*. New York: Harper & Row, 1975.

Lachman, Gary. *A Secret History of Consciousness*. Great Barrington, Mass.: Lindisfarne Books, 2003.

Landis, Leonard. *Psychoanalysis and Beyond*. New York: American Association of Independent Physicians, 1924.

Leahey, T. H. *A History of Psychology*. Englewood Cliffs, N.J.: Prentice Hall, 2000.

Lerner, Eric. "The Big Bang Never Happened." *Discover* (June 1988): 72–78.

Liotti, M. et al. "Brain Responses Associated with Consciousness (Air Hunger)." *Proc. Nat. Acad. Sci.* 98 (February 13, 2001): 2035–2040.

Lombardi, G. "The Michaelson-Morley Experiment." www.drphysics.com, 1997.

Lord, W. *The Titanic*. New York: Henry Holt, 1955.

Luce, Gay Gaer. *Biorhythms in Human and Animal Physiology*. New York: Dover, 1971.

Luria, Alexander. *Higher Cortical Functions in Man*. New York: Basic Books, 1966.

———. *The Working Brain: An Introduction to Neuropsychology*. New York: Basic Books, 1973.

Lyne, William. *Occult Ether Physics*. Lamy, N. Mex.: Creatopia Productions, 1997.

Lynes, Barry. *The Cancer Conspiracy: Betrayal, Collusion, and the Suppression of Alternative Cancer Treatments*. Delmar, N.Y.: Elsmere Press, 2000.

Margolis, Jonathan. *Uri Geller: Magician or Mystic?* London: Orion Publishing Group, 1999.

Marks, D., and R. Kammann. *The Psychology of the Psychic*. Buffalo, N.Y.: Prometheus Books, 1980.

Martin, T. C. *The Inventions, Researches, and Writings of Nikola Tesla*. New York: Electrical Experimenter Publisher, 1893.

Marvin, Frederich R. *Philosophy of Spiritualism and the Pathology and Treatment of Mediomania*. New York: Asa Butts & Co., 1874.

McCausland, Ian. "Anomalies in the History of Relativity." *Journal of Scientific Exploration* 13, no. 2 (1999): 271–90.

McKenna, D., and T. McKenna. "Towards a Holographic Theory of Mind." *MetaScience Quarterly* 1, no. 1 (1979): 80–90.

Mishlove, Jeffrey. *Roots of Consciousness*. New York: Random House, 1975. Also updated at www.mishlove.com.

Mitchell, E., and J. White, eds. *Psychic Explorations*. New York: Putnam, 1971.

Mitchell, E., and J. White, eds. *Psychic Explorations*. New York: G. P. Putnam & Sons, 1974.

Morris, Richard. *The Edges of Science*. New York: Simon & Schuster, 1990.

Musès, Charles, and Arthur Young. *Consciousness and Reality*. New York: Avon Books, 1972.

———. "Imaginary Numbers." In *Consciousness and Reality*. New York: Avon Books, 1972.

Nyberg, Michael. "Killing Mosquitoes with Sound." www.ctsciencefair.org/news/2001.

O'Reagan, Brendan. "The Emergence of Paraphysics." In *Psychic Explorations.* Edited by E. Mitchell and J. White. New York: G. P. Putnam & Sons, 1974.

Ornstein, Robert. *The Psychology of Consciousness.* New York: Viking Press, 1972.

Ouspensky, P. D. *The 4th Way.* New York: Vantage Press, 1957.

———. *In Search of the Miraculous.* New York: Harcourt & Brace, 1960.

———. *New Model of the Universe.* New York: Random House, 1971.

Overbye, Bjorn. "Warped Minds, Bent Truths." *Nexus* 14, nos. 5–6 (2007).

Panati, C. *Supersenses.* New York: Quadrangle Books, 1974.

Penrose, Roger. "On the Origin of Twistor Theory." In *Gravitation and Geometry,* http://users.ox.ac.uk, 2007.

———. *Shadows of the Mind: A Search for the Missing Science of Consciousness.* London: Oxford University Press, 1994.

Persinger, M. A. "ELF Mediation in Spontaneous Psi Events." In *Psychoenergetic Systems* 3 (1979): 155–69.

Pribram, Karl. *Languages of the Brain.* New York: Prentice Hall, 1971.

Price, Lew, and B. Herbert Gibson. "Is There a Dynamic Ether? A New Reality for 21st-Century Physics." www.promedia.net/users/greenbo, 2007.

Progoff, Ira. *Jung, Synchronicity and Human Destiny.* New York: Julian Press, 1973.

Puharich, Andrija. *Beyond Telepathy.* New York: Anchor Press, 1973.

———. *The Iceland Papers.* Ontario: PACE Press, 1979/1996.

———. *Uri: A Journal of the Mystery of Uri Geller.* New York: Doubleday, 1974.

Rampa, Lobsang. *Feeding the Flame.* London: Corgi Books, 1971.

———. *Wisdom of the Ancients.* London: Corgi Books, 1974.

Randall, Lisa. *Warped Passages.* New York: HarperCollins, 2006.

Ratzlaff, John, ed. *Solutions to Tesla's Secrets.* Milbrae, Calif.: Tesla Book Co., 1981.

———. *Tesla Said.* Milbrae, Calif.: Tesla Book Company, 1984.

Read, Herbert. *Origins of Form in Art.* New York: Horizon Press, 1965.

Reich, Wilhelm. *Cosmic Superimposition.* Rangeley, Maine: Orgone Institute Press, 1953.

Revkin, Andrew. "Lisa Randall, Physicist." *Rolling Stone* (November 15, 2007).

Reyburn, Scott. "Uri Buys Spoons in $3.8 Million Savoy Sale at Boham's." www.bloomberg.com, 2007.

Rhine, J. B. *New World of the Mind*. New York: William Sloan Associates, 1953.

Rhodes, Richard. *The Making of the Atomic Bomb*. New York: Simon & Schuster, 1986.

Richardson, K. I. T. *The Gyroscope Applied*. New York: Philosophical Library, 1954.

Roberts, Jane. *Seth Speaks*. Upper Saddle River, N.J.: Prentice Hall, 1972.

Rothman, Milton. *The Laws of Physics*. Greenwich, Conn.: Fawcett, 1963.

Rucker, R. *Geometry, Relativity and the Fourth Dimension*. New York: Dover Publications, 1977.

Rudhyar, Dane. *The Astrology of Personality*. Garden City, New York: Doubleday, 1970.

———. *The Sun Is Also a Star*. New York: Dutton, 1975.

Russell, Bertrand. *Human Knowledge: Its Scope and Limitations*. London: George Allen & Unwin, 1948.

Saltmarsh, H. F. *Foreknowledge*. London: G. Bell & Sons, 1938.

Seifer, Marc. "Co-Evolution of Science and Spirit: Conference report of David Bohm, Roland Fischer, Karl Pribram, and Pir Vilayat Kahn." *MetaScience Quarterly* 4 (October 13, 1987) (unpublished).

———. "Consciousness and the Anthropic Principle." *Journal of Conscientiology* (January 1999): 203–20.

———. "Evolution, PK, and the Group Mind." *Parapsychology Review* 10 (September/October 1979): 5, 24–27.

———. "Gurdjieff." *MetaScience Quarterly* 3 (1980): 348–52.

———. "An Interview with Uri Geller." *MetaScience Quarterly* 3 (Autumn 1980): 335–45.

———. *Inward Journey: From Freud to Gurdjieff*. Kingston, R.I.: Doorway Press, 2003.

———. *Levels of Mind*. Master's thesis. Chicago: University of Chicago, 1974.

———. "The Mind of the Skeptic and the Hierarchy of Doubt." *MetaScience Quarterly* 3 (1980): 285–95.

———. "Parapsychology and Esoteric Thought." *Parapsychology Review* (May–June 1981): 18–21.

———. "The Physics of Consciousness." *Journal of Occult Studies* (August 1977): 148–57.

———. "Precognition, Synchronicity and the Structure of Time." *MetaScience Quarterly* 4 (1981): 76–98 (unpublished). Special Studies, Saybrook Institute, 1982.

———. "Retrocognitive and Precognitive Target Displacements in Card Reading and Dream Telepathy Experiments." *MetaScience Quarterly* 2 (1979): 42–52.

———. "Second Annual International Kirlian Research Association Conference." *MetaScience Quarterly* 1 (Spring 1979): 37–47.

———. *Speculations on the Nature of the Mind.* Unpublished, 1976.

———. "Synchronicity and the Structure of the Psyche." *Journal of Conscientiology* (January 2001): 193–216.

———. "Synchronicity and the Structure of Time." *Consciousness Research Abstracts,* Towards a Science of Consciousness Symposium, University of Arizona (1996): 155–56.

———. "Taking on Einstein." *Extraordinary Science* VIII, no. 1 (January/February/March 1996): 38–43.

———. "The Tunguska Incident." *Tesla Journal* 3rd quarter (1997): 5–6.

———. "The Universe Is a Holarchy." *MetaScience Quarterly* 1 (1979): 92–100.

———. *Wizard: The Life and Times of Nikola Tesla.* New York: Citadel Press, 1997.

Seifer, Marc, and Howard Smukler. "The Puharich Interview." *Gnostica* (September 1978): 21–24, 78–81.

Shapkin, Mikhail. "Unknown Manuscript of Nicola Tesla." http://farshores.org/wmtesla.htm.

Smith, Alpheus, and J. Cooper. *Elements of Physics.* New York: McGraw Hill, 1975.

Smukler, H., and M. Seifer. "Remote Viewing, California to Rhode Island." *MetaScience Quarterly* 1 (1979): 25–29.

Sperry, R. W. "Cerebral Organization and Behavior." *Science* 133 (June 2, 1961): 1749–57.

———. "The Great Cerebral Commissure." *Scientific American* 210, no. 18 (January 1964): 42–52.

Stapp, Henry. *Mind, Matter and Quantum Mechanics*. New York: Springer, 1993.

Steiner, Rudolf. *An Outline of Occult Science*. New York: Anthroposophic Press, 1972.

Stewart, Sharla. "How to Catch a Higgs." *University of Chicago Magazine* (2001): 21.

Taub-Bynum, Edward Bruce. "Psi, Dreams and the Family Unconscious." *MetaScience Quarterly* 4 (1987): 29–45 (unpublished).

Taylor, Alfred. "Meaning and Matter." In *Consciousness and Reality*. Edited by Arthur Young and Charles Musès. New York: Outerbridge/Dutton, 1972.

Taylor, John. *Superminds*. New York: Warner, 1977.

Tesla, Nikola. "The Eternal Source of Energy of the Universe: Origin and Intensity of Cosmic Rays." In *Tesliana*. Belgrade: Tesla Museum, 1932.

———. "The Problem of Increasing Human Energy." *The Century Magazine* (June 1901).

———. "World System of Wireless Transmission of Energy." *Telegraph & Telephone* (October 16, 1927).

Thomsen, Dietrick. "Kaluza-Klein: The Konsiberg Connection." *Science News* (July 7, 1984): 12–14.

Toben, Bob, Jack Sarfatti, and Fred Wolfe. *Space-Time and Beyond*. New York: Dutton, 1985.

Tobin, Paul. *Is the Universe "Fine-Tuned" for Life?* www.geocities.com, 2007.

Tyson, Peter. *The Legacy of $E = MC^2$*. Nova, www.pbs.org, 2005.

Uchii, S. "Mach's Principle." Kyoto University, 2001. www.bun.Kyoto-u.ac.jp, 2001.

Urry, Meg. "The Secrets of Dark Energy." *Parade* (May 27, 2007): 4–5.

Van Flandern, Tom. "The Speed of Gravity." www.Idolphin.org, 2007.

Vasiliev, L. L. *Experiments in Distant Influence*. New York: Dutton, 1976.

Vaughan, Alan. *Patterns of Prophecy*. New York: Doubleday, 1973.

———. "Synchronicity." *Psychic Magazine* (August 1975): 17, 56–61.

Velikovsky, Immanuel. *Worlds in Collision*. New York: Macmillan, 1950.

Wallard, Kathy. "Why Our Planet and Galaxy Take Us for a Spin." *Newsday* (March 9, 2004).

Walsh, B., M. Seifer, and H. Smukler. "Synchronicity and Seriality as a Partial Explanation of the Large Number of Oil Spills in New England." *Journal of Occult Studies* 1 (1977): 76–84.

Watterson, Bill. *The Authoritative Calvin and Hobbes*. New York: Andrews McMeel Publishing, 1990.

Weiss, Peter. "Jiggling the Cosmic Ooze." *Science News* 159 (March 10, 2001): 152–54.

Whicher, O. *Projective Geometry*. London: Steiner Press, 1971.

White, John, ed. *Frontiers of Consciousness*. New York: Avon, 1974.

Whitrow, G. J. *The Nature of Time*. New York: Penguin, 1980.

Whyte, L., ed. *Hierarchical Structures*. New York: Elsevier Publisher, 1969.

Wigner, Eugene. "The Place of Consciousness in Modern Physics." *Consciousness and Reality*. Edited by Musès and Young. New York: Avon Books, 1972.

Wilber, Ken. *Spectrum of Consciousness*. Wheaton, Ill.: Quest, 1977.

Wilcox, David. *Convergence VIII: Scientific Proof of the Nature of a Multi-Dimensional Harmonic Universe*. www.ascension200.com, 2000.

Williams, Dennis A. "The Energy Tangle." *Newsweek* (April 16, 1979): 72.

Williams, L. *The Origins of Field Theory*. New York: Random House, 1966.

Wolf, Fred Alan. *Star Wave: Mind, Consciousness and Quantum Physics*. New York: Macmillan, 1984.

Wolfram, Stephen. "The Nature of Space Notes from a New Kind of Science." www.wolframscience.com, 2003.

Wright, Robert. *Three Scientists and Their Gods: Looking for Meaning in an Age of Information*. New York: Time Books, 1988.

———. "Did the Universe Just Happen?" *Atlantic Monthly* (April 1988): 29–42.

York, Warren. *Scalar Technology*. www.teslatech.info, 2007.

Young, Arthur. *Geometry of Meaning*. New York: Delacorte Press, 1975.

———. *Reflexive Universe*. New York: Delacorte Press, 1973.

Young, Louise B. *The Unfinished Universe*. New York: Touchstone Books, 1987.

INDEX

ABOUT THE AUTHOR

Marc J. Seifer has a bachelor of science degree from the University of Rhode Island and did postgraduate work in graphology at New School University. He earned a master's degree in psychology from the University of Chicago and a doctorate in psychology from Saybrook Institute. He has taught courses on consciousness at Providence College School of Continuing Education for fifteen years and is presently a visiting lecturer in psychology at Roger Williams University.

An expert on the inventor Nikola Tesla and also in the field of graphology, Marc Seifer has lectured at West Point Military Academy, Brandeis University, the Federal Reserve Bank in Boston, the United Nations, CCNY, Lucasfilms Industrial Light & Magic, Oxford University and Cambridge University, and the University of Vancouver, and at conferences in Israel, Serbia, Croatia, and throughout the United States. His articles have appeared in *Wired, Civilization, Parapsychology Review, Journal of Psychohistory, Engineering Dimensions, Consciousness Research Abstracts, Psychiatric Clinics of North America,* and *Cerebrum.* Featured in *Brain/Mind Bulletin, New York Times, Cosmopolitan, Washington Post, New Scientist, The Economist, Rhode Island Monthly,* and on the back cover of Uri Geller's book *MindMedicine,* he has also appeared on national public radio's "To the Best of Our Knowledge" and on the *History Channel* and *Associated Press International TV News,* and been a consultant for *Biography, 60 Minutes,* and *The American Experience.*

Among his other works are *Inward Journey: From Freud to Gurdjieff* (the basis for the first part of this new work), *Framed: A True Courtroom Thriller, The Definitive Book of Handwriting Analysis, Hail to the Chief* (screenplay), *Staretz Encounter: A Psi-Fi Thriller* (novel), and the biography *Wizard: The Life and Times of Nikola Tesla.* Called "a serious piece of scholarship" by *Scientific American,* "revelatory" by *Publisher's Weekly,* and a "masterpiece" by bestselling author Nelson DeMille, *Wizard* is "Highly Recommended" by the American Association for the Advancement of Science.